DATE DUE			

PHOTOCHEMISTRY AND MOLECULAR REACTIONS

M. Mousseron-Canet and J. -C. Mani

PHOTOCHEMISTRY AND MOLECULAR REACTIONS

Translated from French
by J. Schmorak

Israel Program for Scientific Translations
Jerusalem 1972

This book is a translation of
PHOTOCHIMIE ET RÉACTIONS
MOLÉCULAIRES
Dunod
Paris 1969

IPST Cat. No.2318
ISBN 0 7065 1120 4

Printed and Bound in Israel
Printed in Jerusalem by Keter Press
Binding: Wiener Bindery Ltd., Jerusalem

FOREWORD

Photochemistry comprises a frontier imbued with the unknown and as such holds particular fascination. Its possible applications, as yet confined to the synthesis of complex molecules, certain polymerization and halogenation reactions, and the nature of photochromic substances, are still without doubt largely unsuspected. Future projects, concerned for example with air pollution, will open up fields of application beyond the realms of classical technology.

Following the example of other scientific disciplines, photochemistry has benefited from the enormous advances made in spectroscopy: the developments of nuclear magnetic resonance, electron paramagnetic resonance, spectrofluorometry, flash photolysis, which have provided precious information on the nature of excited species, and reinforced by analytical methods adapted for very minute quantities, such as mass spectrometry and gas phase chromatography, have permitted the photochemist to approach problems from their fundamental or their more complex aspects.

During recent years, a number of texts on photochemistry have been published, mostly in the medium of the English language. The state of development of photochemistry in France and the increasing use of our language throughout the world have rendered the publication of a French text desirable. M. Piganiol was the proponent of the idea that a book be published on the subject of "molecular reactions and photochemistry." We offer him our thanks for considering us worthy of fulfilling this task.

Our intention was not only to write a book suitable for organic chemists making their first steps in photochemistry, but also to familiarize them with the recent advances in this science without recourse to an unduly rigorous mathematical treatment.

We have deliberately stressed the fundamental principles of photochemistry, without which no real progress would have been possible. The relationships between the nature of excited states and reactivity are as yet only imperfectly understood, but all studies concerning energy transfers and photosensitization are particularly important, and are certain to yield fruitful results in future.

In writing the book, we chose not to make a compilation of organic photochemical reactions, but rather selected certain subjects which seemed to us particularly important, particularly timely or particularly attractive; an example is the chapter on electrocyclic reactions, recent advances in which provide an interpretation for a large number of organic chemical reactions. We take this opportunity to offer our heartfelt thanks to Professor Woodward (Nobel Prize laureate), who had the kindness to give us some valuable information during his brief visit at Montpellier.

We have dealt at some length with the problem of photooxidation, both because of its obvious preparative interest and of the great variety of partly contradictory views which have been advanced on this subject. We may mention that a very early contribution to this problem was made by a French chemist, Professor Dufraisse.

A large part of future research will probably be devoted to interdisciplinary scientific areas. One of the major future tasks of organic chemistry will be the interpretation of biological mechanisms, in which spectroscopy will be of immense assistance. We have stressed the contribution of fluorescence to the study of complex macromolecules (e.g., the use of hydrophobic fluorescent detectors), and have also discussed the biological implications of charge transfer complexes. Certain aspects of photobiology have been reviewed; the recent theories of Professor Rosenberg concerning photoconductance during the visual process were considered especially interesting.

The last part of the book contains a brief introduction to experimental methods: radiation sources and selection of types of radiation, actinometric methods, degassing of solutions, techniques of flash photolysis and the determination of quantum yields. For a detailed account of these subjects, the reader is referred to the exhaustive treatise by Calvert and Pitts.

At the end of each chapter a short list of fundamental texts or available reviews is given, rather than a detailed bibliography.

The Equipe de Photochimie du C.N.R.S.* was organized in Montpellier, after one of us (M.M.C.) had spent a certain period of time in the laboratories of Professor Barton, whose synthesis of aldosterone was a striking illustration of the potentialities of photochemistry in the synthesis of molecules of complex structure.

This book could not have been written without the knowledge gained by one of us (J.C.M.) on the subject of mechanistic interpretation of photochemical phenomena during a period spent in the laboratories of Professor Pitts.

Our thanks are due to Professor Mousseron, who kindly put at our disposal the text of some of his lectures in photochemistry and encouraged us in our task with his customary enthusiasm. We also thank Mr. F. Mani for his valuable assistance in the compilation of the index. We must also mention our laboratory team, since it is in their company that we study photochemistry in the course of our weekly seminars. We also thank Mr. J. Grasset for his help in typing the manuscript of the book.

Magdeleine Mousseron-Canet
Jean-Claude Mani
Equipe de Photochimie du C.N.R.S.
Montpellier, Ecole Nationale Supérieure de Chimie
June 1968.

* [Photochemistry Section of the French National Science Research Council (Centre National de la Recherche Scientifique).]

CONTENTS

Chapter 4

Chapter 5

Chapter 6

Chapter 1

DEFINITIONS. LAWS OF PHOTOCHEMISTRY

I. ABSORPTION OF LIGHT

A photochemical reaction is produced by the absorption of electromagnetic radiation by a molecule. According to Planck's quantum theory, this energy absorption takes place by steps, each step or transition corresponding to the absorption of a "quantum" of energy (photon). The energy E of this quantum is given by Planck's equation:

$$E = h\nu \tag{1.1}$$

where h is Planck's constant and ν is the frequency of the absorbed radiation. Planck's equation can also be written as follows:

$$E = hc/\lambda \tag{1.2}$$

where c is the velocity of light and λ is the wavelength of the radiation. The energy calculated by Planck's equation (1.1 or 1.2) is expressed in ergs. It may be calculated in kcal/mole by the formula

$$E(\text{kcal}\cdot\text{einstein}^{-1}) = \frac{2.859\,1 \times 10^5}{\lambda(\text{Å})}. \tag{1.3}$$

Table 1.1 gives the values of the different constants and the definitions of a number of units.

TABLE 1.1

h (Planck's constant) = 6.6256×10^{-27} erg·sec·quantum^{-1}
c (velocity of light) = 2.9979×10^{10} cm·sec^{-1}
N (Avogadro's number) = 6.023×10^{23}
Wave number (cm^{-1}) : $\omega = 1/\lambda$ (λ in cm)
1 einstein = N quanta

It is important to note that the absorption of light of a wavelength shorter than 7500 Å (visible and UV) supplies an amount of energy which is sufficient to effect electronic transitions. In practice, the photochemist will work with wavelengths between 1850 and 7500 Å; the energy supplied by waves longer than 7500 Å is usually too small to produce an electronic transition, while energies corresponding to wavelengths below 1850 Å are too large and will produce dissociation of the molecules.

II. ELECTRONIC TRANSITIONS. EXCITED STATES

An electronic orbital is a volume element in space, in which the probability of finding an electron is very large (99.9%); it is calculated from the wave function* of a single electron and is assumed to be independent of all other electrons in the molecule.

An electronic state is determined by all the electrons in all the orbitals; the wave function of an electronic state is a combination

* Without going into details of wave mechanics, we may say that all electronic states of a molecule may be described by wave functions ψ which represent solutions of Schroedinger's equation. The molecular wave function defines the orbits and the properties of the electrons of the molecule.

of the wave functions of each electron in each of the orbitals in the molecule. If, in a molecule, an electron changes its orbital, the electronic state of the molecule is changed as a result; it is then necessary to consider the states of the molecule and not merely the orbitals, since the inter-electron effects are very important.

An electronic transition is the passage of an electron from one orbital to another.

The ground state of a molecule is its normal state and corresponds to the minimum energy. Each molecule has only one ground state.

The absorption of energy by a molecule results in three types of transitions: 1) electronic transitions; 2) vibrational transitions; 3) rotational transitions.

Like electronic transitions, the rotations and vibrations of a molecule are quantized. The energies corresponding to electronic transitions are between 40 and 250 kcal $\cdot$ mole^{-1}, while those of vibrational and rotational transitions are much smaller – 1–10 kcal$\cdot$mole^{-1} and 0–1 kcal$\cdot$mole^{-1} respectively. It follows that, for each electronic level of a molecule, there are several vibrational sub-levels, and for each vibrational level there are several rotational sub-levels.

The electronic excited state (or, as it is often called, the excited state) of a molecule is a state resulting from an electronic transition. Figure 1.1 is a schematic representation of potential energies of a molecule.

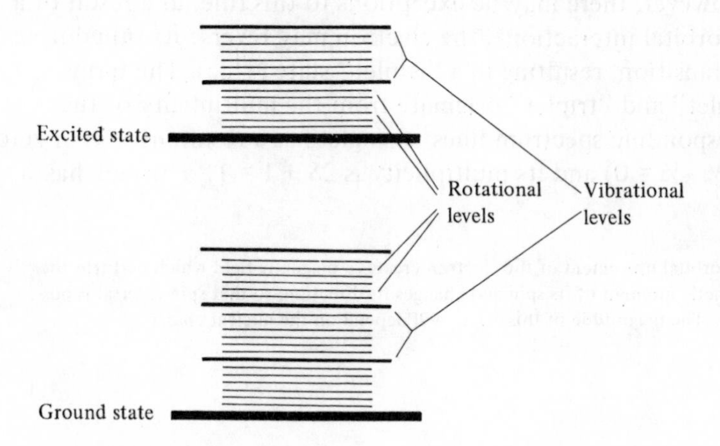

In most organic molecules in their ground state all electrons have anti-parallel spins: this is the so-called "singlet state." Figure 1.2 represents a molecular orbital: when the molecule is in the ground state, the ground orbital is occupied by two anti-parallel electrons, while the orbital corresponding to the higher energy level is vacant (1.2a). During the transition, one of the electrons will pass into this vacant orbital. Wigner's spin conservation principle, according to which the resultant spin momentum of the system must remain constant, indicates that the spin must be conserved during the transition, and the excited state will also be a singlet state (1.2b).

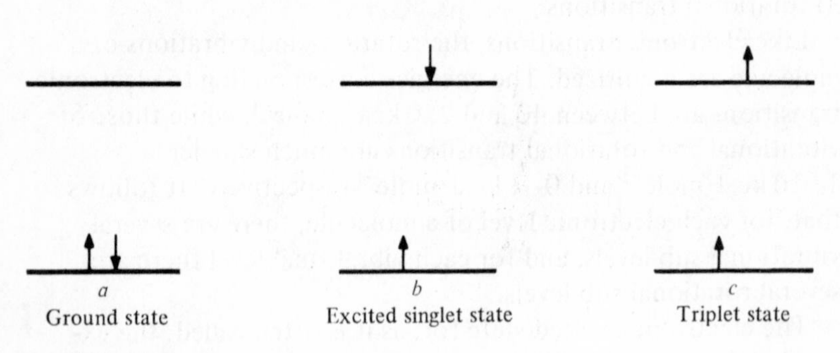

FIGURE 1.2

However, there may be exceptions to this rule: as a result of a spin-orbital interaction*, the electron may reverse its spin during the transition, resulting in a "triplet" state (1.2c). The terms "singlet" and "triplet" originate from the multiplicity of the corresponding spectrum lines: a singlet has a resultant spin of zero ($S = ½ - ½ = 0$) and its multiplicity is $2S + 1 = 1$; a triplet has a

* The orbital movement of the electron creates a magnetic field which perturbs the magnetic moment of its spin and changes its direction, so that spin reversal is possible. The magnitude of this effect will depend on the nuclear charge.

resultant spin of unity* ($S = \frac{1}{2} + \frac{1}{2} = 1$), and its multiplicity is $2S + 1 = 3$.

Transition moment. The transition moment is a vector, which may be calculated from wave-mechanical wave functions of the two states between which the transition takes place, and which is a characteristic of this transition. An example of the application of this moment will be given in Chapter 12, in connection with fluorescence polarization.

Bonding and antibonding orbitals. In wave mechanics, the wave functions of two atoms which are bound to each other can be combined in two different manners; there result two molecular orbitals with different electron distributions. Figure 1.3a represents a "bonding," while Figure 1.3b represents an "antibonding" orbital. In an antibonding orbital, the probability of finding an electron between the two nuclei is very small. In general, low electronic levels are bonding, while the higher levels are antibonding.

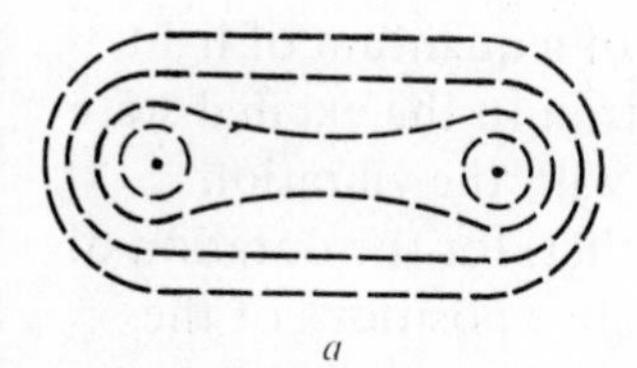

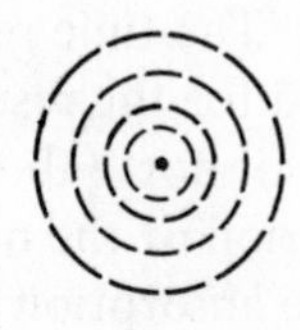

a b

FIGURE 1.3

So-called "nonbonding" orbitals correspond to the unshared electrons in an atom (e.g., in oxygen or nitrogen).

Nomenclature of electronic states. As a rule, we shall utilize the nomenclature of molecular orbitals. The orbitals of organic molecules are generally σ, π, n, π^*, σ^*; the antibonding orbitals are marked with an asterisk (π^*, σ^*); n is a nonbonding orbital.

* Whereas a singlet state is diamagnetic ($S = 0$), a triplet state ($S = 1$) is paramagnetic and, because of its magnetic properties, can be studied by EPR spectroscopy.

In order to represent an electronic transition, the upper state is written first, while the lower state is written last, the direction of the transition being indicated by an arrow. If, for instance, we wish to indicate the transition of an electron from a bonding orbital π to an antibonding orbital π^*, we shall write $\pi^* \leftarrow \pi$, and the excited state which corresponds to this transition will be written $(\pi^* \leftarrow \pi)$. In practice, $\pi \rightarrow \pi^*$ and $(\pi \rightarrow \pi^*)$ are also frequently written.

Other types of nomenclature of the excited states may also be found in the literature, of which the most important one is the symmetry notation. The state in question is described according to the behavior of the wave function in the symmetry operations of the group to which the molecule belongs. For an example of this type of notation see benzene, in Chapter 5.

III. THE FRANCK-CONDON PRINCIPLE

"The time required for the absorption of a quantum of light and for the resulting transition of an electron to the excited state is so short (about 10^{-15} sec) as compared with the vibration period of the molecule (about 10^{-13} sec) that, for the duration of the absorption and the excitation, the relative positions of the nuclei (internuclear distance r) and their kinetic energies do not change significantly."

It follows directly from this principle that electronic transitions between two potential energy surfaces can be represented by vertical lines which join these surfaces (Figure 1.4). A "vertical transition" is a transition which obeys the Franck-Condon principle.

Figure 1.4 shows the potential energy curves of a molecule AB in its ground state and in its excited state. The internuclear distance r_0 corresponds to the zero vibrational level of the ground state from which the transition occurs. As a result of a vertical transition, the excited molecule will attain the vibrational level ν, for which $r' = r_0$; if the geometry of the excited state is not identical with that of the ground state , which is generally the case,

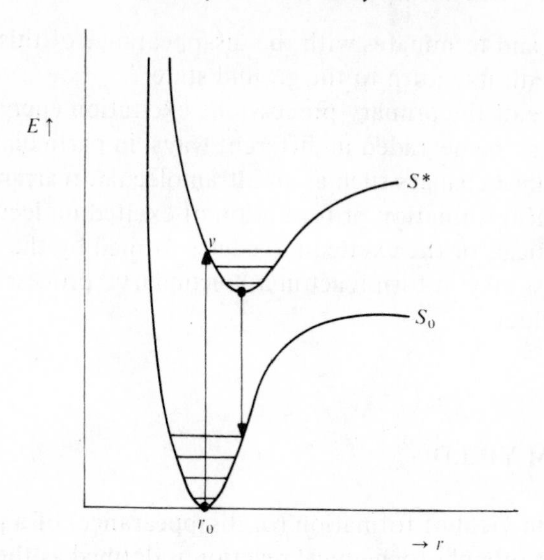

FIGURE 1.4

this vibrational level will not be zero. Thus, the absorption maximum does not as a rule correspond to a 0–0 transition (which is the transition between two nonexcited vibrational states).

The total energy received by the molecule which has attained the vibrational level ν of the excited state may well exceed its dissociation energy; nevertheless, the molecule can remain for 10^{-8} seconds in the excited state and then return to the ground state without undergoing dissociation. This indicates that the energy of the excited electron cannot be instantaneously converted to the kinetic energy of the nuclei, unless another change also takes place (pre-dissociation or internal conversion).

IV. PRIMARY AND SECONDARY PROCESSES

According to Noyes, a primary photochemical process is "a succession of events which begins with the absorption of a photon by a molecule and terminates with the disappearance of this molecule, or with its return to the ground state."

In the course of the primary process, the excitation energy of the electron may be degraded in different ways, in particular owing to chemical changes such as an intramolecular rearrangement, free radical formation or formation of excited molecules.

The free radicals or the excited molecules formed by the primary process may in turn react in a "secondary" process to yield new products.

V. QUANTUM YIELD

The quantum yield of formation (or disappearance) of a product as a result of a photochemical reaction is defined as the number of molecules of this product which are formed (or which disappear) per quantum of light absorbed.

It is highly important to distinguish between the primary quantum yield φ and the overall quantum yield Φ. The former is the quantum yield of the product formed as a result of the primary process, while the latter is calculated by determining the total number of molecules formed at the end of the reaction, i.e., it includes products formed in secondary processes. Let a molecule M undergo a photochemical reaction to form the radical M· in the primary process; let this radical then react to yield products A, B, C... . The primary quantum yield of M· will then be defined by the relationship

$$\varphi_{M\cdot} = \frac{d[M\cdot]/dt}{I_a} \qquad (1.4)$$

where I_a is the radiation intensity absorbed by M, i.e.:

$$\varphi_{M\cdot} = \frac{\text{Number of M· radicals formed per cm}^3\text{ per second}}{\text{Number of quanta absorbed by M per cm}^3\text{ per second}}\,.$$

The quantum yield of the final product A will be

$$\Phi_{\mathrm{A}} = \frac{\mathrm{d}[A]/\mathrm{d}t}{I_a} = \frac{\text{Number of A molecules formed per cm}^3\text{ per second}}{\text{Number of quanta absorbed by M per cm}^3\text{ per second}} .$$

VI. FUNDAMENTAL LAWS OF PHOTOCHEMISTRY

First Law (Grotthus, 1817; Draper, 1843):

The light absorbed by a molecule is the only light which can produce photochemical changes in the molecule.

Second Law (Stark-Bodenstein, 1913):

The absorption of light by a molecule is a one-quantum process, in which the sum total of primary quantum yields must be equal to unity

$$\Sigma\varphi = 1 .$$

Since many primary processes result in the destruction or deactivation of the excited molecule (chemical reactions, radiative or radiationless deactivations, energy transfer), the sum total of the primary quantum yields of a product will always be less than or equal to unity. This does not apply to the overall quantum yield, which may be much larger than unity; thus, for certain chain reactions, it may be as high as 10^4.

It is important to realize that the Second Law is only valid for classical photochemical reactions, in which the concentration of the excited molecules formed is very low; under these conditions the absorption of a second light quantum by the excited molecule is extremely improbable. Flash photolysis may involve two-photon processes: a very powerful flashlight or a laser results in a fairly high concentration of excited molecules; a second flash may result in the absorption of a second quantum by a few excited molecules (for a brief review of the relevant experimental techniques see Chapter 14).

Bibliography

1. Calvert, J. G. and J. N. Pitts, Jr. Photochemistry. John Wiley and Sons, Inc., N. Y. 1966.
2. Pitts, Jr., J. N., F. Wilkinson, and G. S. Hammond. Advances in Photochemistry. Edited by W.A. Noyes, Jr., G. S. Hammond, and J. N. Pitts, Jr., 1 : 1–22. Interscience Publishers, N. Y. 1963.

Chapter 2

ELECTRONIC TRANSITIONS

I. ABSORPTION OF LIGHT

The absorption of a photon by a molecule results in a change in its electronic level (Chapter 1). An absorption spectrum is characterized by the wavelength of maximum absorption and by the intensity of this absorption, which is generally reported as the molar extinction coefficient ϵ. The absorption wavelength is a measure of the electronic energy level attained by the molecule, whereas the intensity of the absorption band is a measure of the probability of electronic transition.

1. Absorption wavelength

Were it not for the existence of rotational and vibrational sub-levels, a transition between two levels would yield an absorption spectrum of a single line, the wavelength of which would be an accurate indication of the energy difference between the two levels (equation 1.3). The presence of vibrational and rotational sub-levels is responsible for the complicated aspect of the absorption spectra, which most often appear as wide bands; the exact identification of the 0–0 band (Chapter 1, Section III) is difficult. The maximum absorption wavelength is not an exact measure of the relative energies of the two levels, since, in accordance with the Franck-Condon principle, it does not, as a rule, correspond to the 0–0 transition. Nevertheless, the spectra of certain compounds, in particular aromatic compounds, have a fine structure due to 0–0, 0–1, 0–2 ... transitions between the ground state and the successive vibrational levels of the excited state.

2. Intensity of absorption bands

The intensity of an absorption band, which is a measure of the probability f of the electronic transition, can be determined experimentally by integrating the absorption spectrum. The probability of the transition involves three different factors:

a) a factor P_m which concerns the relative multiplicities of the levels between which the transition occurs;

b) a factor P_0, which is a measure of the degree of overlap of the orbitals involved;

c) a factor P_s, which concerns the symmetry of the wave functions of the ground state and of the excited state.

The transition probability is the product of these three factors:

$$f = P_m . P_0 . P_s .$$

If the two levels are of the same multiplicity, $P_m = 1$. If the multiplicity changes in the course of the transition, $P_m = 10^{-5} - 10^{-6}$. Clearly, the probability of a singlet → triplet or a triplet → singlet transition will be very small; these are the so-called "forbidden" transitions.

The value of P_0 varies between unity and 10^{-4}, depending on the position of the orbitals. We shall see that this factor is relevant in differentiating $n \rightarrow \pi^*$ transitions from $\pi \rightarrow \pi^*$ transitions.

The parameter P_s varies between unity and 10^{-3}. It is of considerable importance, mainly in the case of symmetrical aromatic molecules and polyenes. The value of P_s will be small if the molecule loses its symmetry on passing from the ground to the excited state. We shall return to this subject in connection with symmetry notation (Chapter 5).

The intensity of the absorption band indicates whether a transition is allowed or forbidden. The lifetime τ of the excited state is inversely proportional to the band intensity I:

$$\tau = \frac{3.5 \times 10^8}{\omega_m^2 I} \tag{2.1}$$

where ω_m is the average absorption frequency in cm^{-1} and $I = \int \epsilon \, d\omega$, where ϵ is the molar extinction coefficient. If the absorption curve is symmetrical, we have $I = \epsilon_{max} \cdot \Delta\omega_{1/2}$, where ϵ_{max} is the maximum molar extinction coefficient and $\Delta\omega_{1/2}$ is the width (in cm^{-1}) of the absorption band between half-intensity points.

II. JABLONSKY'S DIAGRAM

Jablonsky's diagram (Figure 2.1) is a schematic representation of the redistribution of energy resulting from the absorption of a photon by a molecule.

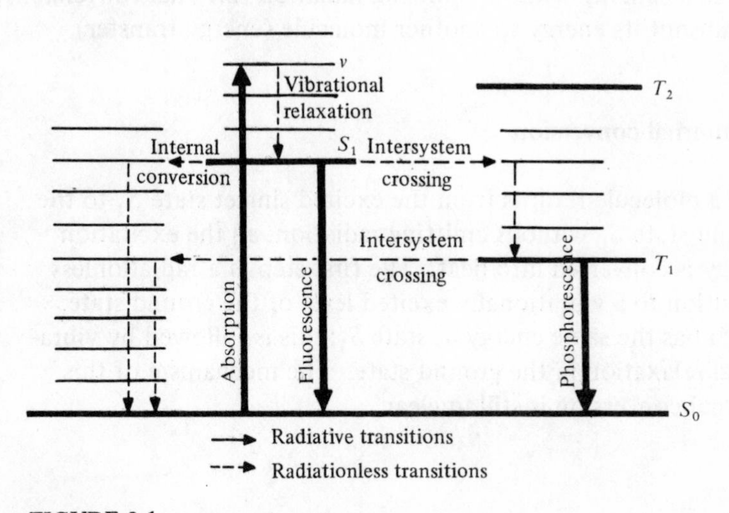

FIGURE 2.1

1. Vibrational relaxation

We shall consider a molecule which has passed, by absorbing a photon, to an excited singlet state S_1 or, more correctly, to the vibrational sub-level ν of state S_1. The excited molecule can now enter a chemical reaction, or can lose its energy by emission of a photon from the vibrational energy level it has attained by the absorption, or else can lose its vibrational energy by a collision and attain the vibrationally nonexcited state S_1. A photochemical reaction will only take place if its rate is faster than the rates of the physical deactivation processes. In the vapor phase and under low pressures, the molecule will lose its energy by reemitting a photon from the vibrationally excited state S_1. In the liquid phase, on the contrary, vibrational relaxation takes place: the molecule loses its excess vibrational energy by collision with the molecules of the solvent or with other molecules of the solute. In solution, this vibrational relaxation process is highly efficient, and takes place within $10^{-13} - 10^{-11}$ sec; as a result, all light emission in solution will take place from the zero vibrational level of the excited state.

The state of the molecule is now $S_1(\nu = 0)$, and the molecule can:

lose its energy by emitting a photon (fluorescence);
be converted to a triplet state by intersystem crossing;
undergo a photochemical reaction;
lose its energy without emitting radiation (internal conversion);
transmit its energy to another molecule (energy transfer).

2. Internal conversion

If a molecule returns from the excited singlet state S_1 to the ground state S_0 without emitting radiation, all the excitation energy is converted into heat. The first step is a radiationless transition to a vibrationally excited level of the ground state, which has the same energy as state S_1; this is followed by vibrational relaxation in the ground state. The mechanism of this internal conversion is still unclear.

If the molecule is excited not to the first excited singlet state S_1, but to the second excited singlet state S_2, it may be assumed, as a rule, that the energy difference between S_1 and S_2 levels is not very large; there is then overlapping of vibrational sub-levels of S_1 and S_2, and an internal energy conversion from S_2 to S_1 which is so efficient that, for all practical purposes, the molecule might have been excited to the level S_1. Hence the following empirical rule: "It may be assumed that the duration of the internal conversion of a molecule to the lowest vibrational state of its lowest excited singlet state is short (about 10^{-12} sec) as compared with the duration of emission of a photon, whatever its original excited singlet state." This rule does not apply if the difference between the lowest excited singlet level S_1 and the level actually attained by excitation S_2 is large; in such cases a photon may be emitted from S_2.

The energy difference between the lowest excited singlet state S_1 and the ground state S_0 is generally large, and the internal conversion $S_1 \rightsquigarrow S_0$ will generally be inefficient.

Temperature is the most important external factor, since a temperature rise increases the number of collisions, thus favoring internal conversion.

3. Intersystem crossing

A triplet state is the result of an excitation accompanied by a reversal of the spin of an electron (Chapter 1); its energy will be smaller than that of the corresponding excited singlet state, since a part of the energy has been used up in flipping the electron spin (Hund's rule).

The triplet state T may be populated in three ways:

1) by direct absorption $S_0 \rightarrow T$. The probability of such an event is very low ($P_m = 10^{-5} - 10^{-6}$), and the event will only occur in the presence of heavy atoms, which increase the energy of a spin-orbit interaction (Chapter 1) or under very particular experimental conditions (e.g., under a very high pressure of oxygen);

2) by energy transfer (Chapter 3);

3) by intersystem crossing, which is a radiationless transition from an excited singlet state to a triplet state.

The mechanism of this crossing is the following. Initially there is coupling between the vibrationally nonexcited state S_1 and the isoenergetic vibrational triplet state; this is followed by vibrational relaxation to vibrational sublevel zero of the triplet state. Thus, intersystem crossing is an internal conversion with a change in the direction of spin and is forbidden by the Wigner rule. It will occur only as a result of spin-orbit interactions, which may change the direction of the spin.

A very important consequence of the spin-orbit interaction is that, in organic molecules, the spins of the two electrons will not be exactly parallel or exactly antiparallel, and there will be no "pure" singlet or triplet states; a singlet state will always be contaminated by some triplet character and vice versa. The mixing coefficient will be proportional to the energy of spin-orbit interaction and inversely proportional to the energy difference between the singlet and triplet states involved.

The probability of an intersystem crossing will depend on several factors:

1) the probability will be higher, the more singlet-like is the triplet state of the molecule, i.e., the larger the energy of spin-orbit interaction and smaller the energy difference between the triplet and the singlet;

2) "the heavy atom effect": the intensity of the spin-orbit interaction is a function of the nuclear charge. In the presence of heavy atoms these interactions are very considerable and the intersystem crossing is facilitated;

3) the smaller the energy difference between the excited singlet and the triplet, the greater the probability of the intersystem crossing. This is because the triplet will then be more singlet-like, and because the electronic configurations of the isoenergetic levels of the triplet and the singlet states will be similar owing to the fact that the vibrational energy of the triplet will be small; thus, the isoenergetic coupling will be facilitated;

4) the presence of paramagnetic substances, i.e., substances with unpaired electron spins, will also increase the probability of this crossing.

In general, the rate of the intersystem crossing will be sufficiently fast to compete with other processes of deactivation of the excited singlet state.

Since the transition triplet–ground-state singlet is forbidden, the lifetime of the triplet state will be much longer than that of the excited singlet – between 10^{-4} and 10 seconds. At room temperature, the molecule in the triplet state will undergo numerous collisions with the solvent molecules; this process of deactivation by collisional transfer (Chapter 3) will be much more intensive than radiative deactivation (phosphorescence) or radiationless deactivation (intersystem crossing). The intersystem crossing $T \rightsquigarrow S_0$ will be the more significant, the smaller the energy difference $T - S_0$; this process is more efficient than the internal conversion $S_1 \rightsquigarrow S_0$.

Photochemical processes do not necessarily involve the lowest triplet state of the molecule; recent studies have shown that higher energy triplet states may also participate in such processes (see Chapter 3, triplet–triplet energy transfer).

III. LUMINESCENCE

1. Fluorescence

After a molecule has reached the lowest vibrational level of an excited singlet state, it can return to its ground state with the emission of a photon. The name of this effect is fluorescence, and the process may be described as follows:

$$S_1 \rightarrow S_0' + h\nu' .$$

In accordance with Planck's equation, the wavelength of the emitted photon will be given by the relationship

$$\lambda' = \frac{hc}{E_1 - E_0'}$$

where E_1 is the energy of the lowest vibrational level of an excited singlet state, and E'_0 is the energy of the vibrational level of the ground state attained by the transition.

If E_0 is the energy of the lowest vibrational level of the ground state, we have, in all cases

$$E'_0 > E_0\,, \quad \text{i.e.,} \quad E_1 - E'_0 < E_1 - E_0\,.$$

If λ_0 is the wavelength of the 0–0 fluorescence band, we shall have

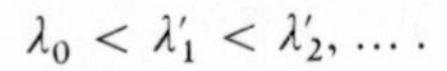

$$\lambda_0 = \frac{hc}{E_1 - E_0} \quad \text{and} \quad \lambda' > \lambda_0\,.$$

In the emission of fluorescence, the initial state is the lowest vibrational level of the excited state S_1, while the final states are the different vibrational levels of the ground state S_0; this results in a series of emission bands of wavelengths λ_0, λ'_1, λ'_2, ..., where

$$\lambda_0 < \lambda'_1 < \lambda'_2, \dots\,.$$

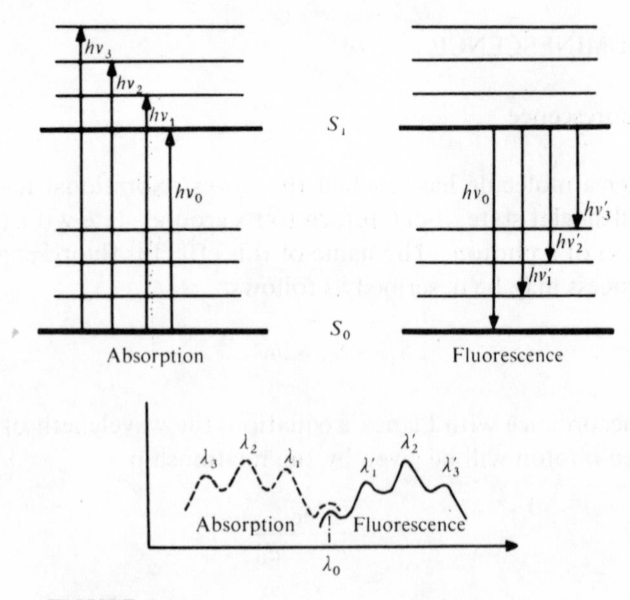

FIGURE 2.2

During the absorption process, on the contrary, the initial state is the lowest vibrational level of the ground state S_0, while the final states are the different vibrational levels of the excited singlet state S_1, which results in a series of absorption bands with wavelengths $\lambda_0, \lambda_1, \lambda_2, \ldots$, where

$$\lambda_0 > \lambda_1 > \lambda_2 \ldots .$$

The band of wavelength $\lambda_0(0\text{–}0)$ is common to fluorescence and to absorption. The fluorescence spectrum and the absorption spectrum are symmetrical about the 0–0 band. The lifetime of the excited singlet state is very short – 10^{-9} to 10^{-5} seconds; the time of quenching of the fluorescence will be of the same order.

The quantum yield of fluorescence can be experimentally determined, but the technique which must be employed is difficult. Some molecules have a fluorescence quantum yield which is practically equal to unity (e.g., fluorescein): all the energy absorbed by the molecule is reemitted as fluorescence. In the case of other molecules, this yield may be practically nil, and the entire energy which has been absorbed is expended in various other processes (internal conversion, intersystem crossing, energy transfer, photochemical reaction).

2. Phosphorescence

After a molecule has attained the lowest vibrational level of the triplet state, it may emit a photon and return to one of the vibrational levels of the ground state. This effect is known as phosphorescence:

$$T_1 \rightarrow S_0 + h\nu'' .$$

The phosphorescence wavelengths will be markedly longer than the absorption wavelengths, since the energy of a triplet is smaller than that of the excited singlet. The 0–0 phosphorescence band, i.e., the band situated at the shortest wavelength, is an experimental measure of the energy of the triplet state. The quenching time of the phosphorescence is of the same order as the lifetime of the triplet state (10^{-4}–10 seconds).

At ordinary temperatures, phosphorescence is in general not observed, since the deactivation of the triplet state is mostly effected by a collisional mechanism. In order for it to be perceptible, it must be measured at very low temperatures, so as to reduce as much as possible the number of collisions. Such work is usually carried out in vitreous solvents at the temperature of liquid nitrogen (77° K).

The relative intensities of fluorescence and phosphorescence processes will depend mostly on the tendency to undergo intersystem crossing. The presence of heavy atoms facilitates this transition and favors phosphorescence at the expense of fluorescence. This fact can be used experimentally in order to intensify the phosphorescence of a molecule: the solvent employed may contain a heavy atom, or else a substance with a heavy atom is introduced as an additive to the solvent. In this way it is possible to observe phosphorescence effects which could not otherwise be detected.

3. Other types of fluorescence

a) *"Excimeric" fluorescence.* An "excimer," or an excited dimer, is a complex of a molecule in the excited singlet state S_1 with a molecule in the ground state S_0. It will only be formed if the concentration of the molecules in the solvent is sufficiently high; such a formation is infrequent, but may occur in certain aromatic hydrocarbons (benzene, toluene).

This excited complex may lose its energy and revert to the ground state, with emission of a photon ("excimeric fluorescence"):

$$S_0 + h\nu \rightarrow S_1 ,$$

$$S_1 + S_0 \rightarrow (S_0 S_1)^* ,$$

$$(S_0 S_1)^* \rightarrow S_0 + S_0 + h\nu' .$$

The wavelength of this fluorescence is much longer than that of conventional fluorescence. Its quantum yield is proportional to the concentration of the molecules.

b) *Delayed type E fluorescence.* Very rarely, a $T \rightsquigarrow S_1$ transition may be observed. The emission of a photon from a state S_1 resulting from this type of transition is known as type E fluorescence, and was first observed for eosin. The term "alpha-phosphorescence" is sometimes also used in the literature, but the effect is really a fluorescence. Since the intersystem crossing is caused by the absorption of thermal energy of the surrounding molecules by the triplet, the intensity of type E fluorescence increases with the temperature:

$$S_0 + h\nu \rightarrow S_1$$

$$S_1 \rightsquigarrow T_1$$

$$T_1 + kT \rightsquigarrow S_1 \quad \text{(thermal excitation)}$$

$$S_1 \rightarrow S_0 + h\nu' .$$

c) *Delayed type P fluorescence (two-photon process).* Delayed type P fluorescence was first noted in the case of pyrene. It has the same wavelength as natural fluorescence and originates from the excited singlet state S_1, but has a much longer lifetime, almost as long as that of phosphorescence. The following mechanism has been proposed to explain this phenomenon:

$$S_0 + h\nu \rightarrow S_1$$

$$S_1 \rightsquigarrow T_1$$

$$T_1 + T_1 \rightarrow S_1 + S_0$$

$$S_1 \rightarrow S_0 + h\nu' .$$

If the concentration of the triplet T_1 is sufficiently high, energy transfer may take place between two triplets to give one excited singlet state S_1 and one ground state S_0. This mechanism, which explains the delayed fluorescence of many organic molecules, is a two-photon process: since one excited molecule receives energy from another excited molecule, at most one photon can be emitted for every two photons absorbed. The intensity of the emission is proportional to the square of the absorbed radiation intensity, because the formation of S_1 is proportional to the square of the concentration of T_1.

In certain cases, an excimer $(S_1S_0)^*$ is formed as an intermediate product; in such a case we have simultaneous type P fluorescence and excimer fluorescence.

IV. THE EXCITED STATES ($\pi \rightarrow \pi^*$) AND ($n \rightarrow \pi^*$)

The two types of electronic transitions which are the most important in organic photochemistry are $\pi \rightarrow \pi^*$ transitions, which are involved in the excitation of double bonds, and $n \rightarrow \pi^*$ transitions, which are typical in the photochemistry of the carbonyl group.

Five types of molecular orbitals are of importance in organic photochemistry: the bonding π orbital, the antibonding π^* orbital, the nonbonding n orbital, the bonding σ orbital, and the antibonding σ^* orbital. Single bonds between two atoms involve σ orbitals: the σ-electrons in these orbitals are strongly bonding and are highly localized; their excitation energy will therefore be an important parameter. Double bonds involve σ orbitals and π orbitals at the same time; the π-electrons are delocalized and are readily excited. Finally, certain hetero atoms such as oxygen or nitrogen contain unshared electrons in n orbitals, which are localized on the hetero atom; these nonbonding n orbitals obviously have no corresponding antibonding orbital.

1. The C=C double bond

Each of the two carbon atoms forming a double bond has an sp^2 configuration: three sp orbitals in the same plane at an angle of 120° to each other and a p orbital which is perpendicular to this plane. Two *sp* orbitals combine to form one bonding orbital σ and one antibonding orbital σ^*. The two *p* orbitals combine to form one bonding π orbital and one antibonding π^* orbital.

Figure 2.3 shows the formation of molecular orbitals from atomic orbitals. The formation of a bonding orbital corresponds to a decrease in energy, while the formation of an antibonding orbital corresponds to an increase in energy.

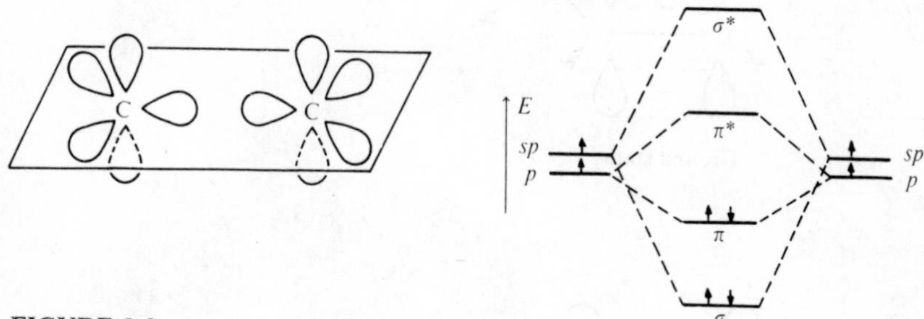

FIGURE 2.3

In the ground state, only the bonding orbitals σ and π are occupied, since there are only four electrons to be placed. The σ and σ^* orbitals do not play a significant part in the electronic transitions encountered in classical photochemistry, since the energies involved are too large; the only transitions of interest are thus $\pi \to \pi^*$, i.e., the transition of an electron from the π orbital to the antibonding π^* orbital.

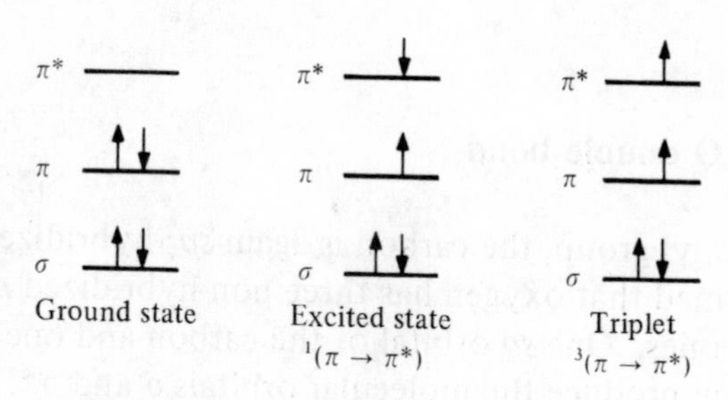

FIGURE 2.4

In the ground state the π orbital is bonding, the electron density between the carbon atoms is high, and the coplanar structure of the molecule is stabilized. In the excited state ($\pi \rightarrow \pi^*$), the electron density between the carbon atoms is much lower, since the π^* orbital is antibonding, and the most stable configuration corresponds to a 90° rotation of the C–C bond (in the case of ethylene).

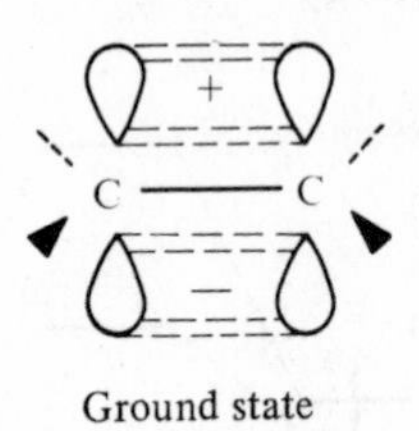

Excited state

FIGURE 2.5. The signs + and – on the two lobes of the orbital refer to the signs of the corresponding wave functions and must not be confused with electric charges. The π and π^* orbitals are delocalized on two or more atomic nuclei.

2. The C=O double bond

In a carbonyl group, the carbon is again sp^2-hybridized, and it may be assumed that oxygen has three non-hybridized *p* orbitals of equal energies. One *sp* orbital of the carbon and one *p* orbital of the oxygen produce the molecular orbitals σ and σ^*. The *p* orbital of the carbon and another *p* orbital of the oxygen form the

π and π^* orbitals. The last p orbital of the oxygen atom remains free, and its energy remains constant: this is the n orbital of the carbonyl.

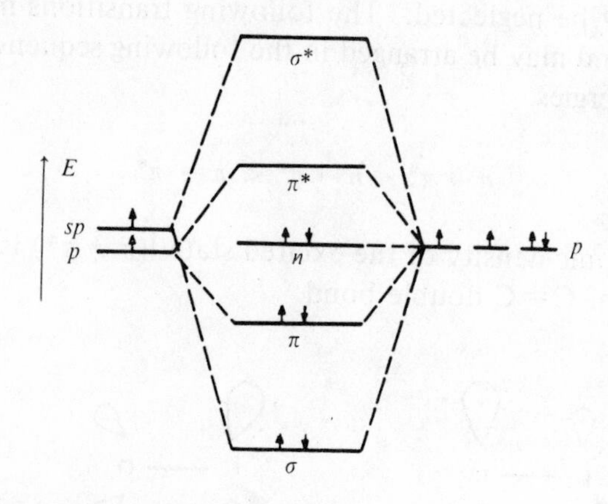

FIGURE 2.6

The magnitudes of the energies of orbitals σ, σ^*, π and π^* are similar for the carbonyl group and the C=C bond.

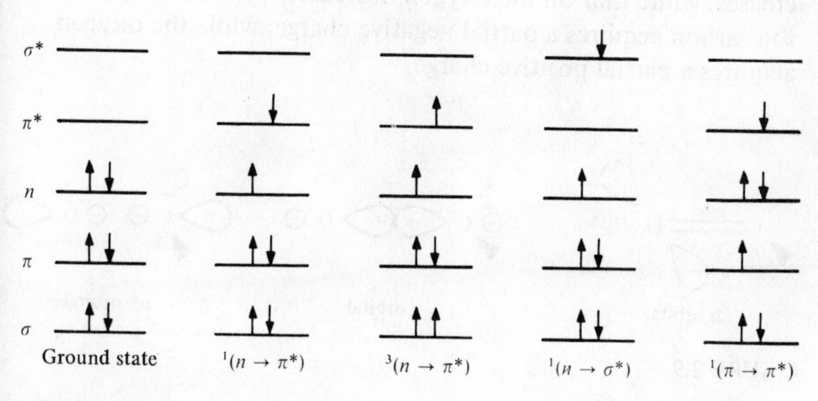

FIGURE 2.7

In the ground state, the occupied orbitals are σ, π and n. Since the energy of orbital n is relatively high the $n \rightarrow \sigma^*$ transition is allowed and actually requires less energy than a $\pi \rightarrow \pi^*$ transition. As in the case of a C=C double bond, transitions from the σ orbital may be neglected. The following transitions must be considered, and may be arranged in the following sequence of increasing energies:

$$n \rightarrow \pi^* < n \rightarrow \sigma^* < \pi \rightarrow \pi^* .$$

The electronic density of the excited state $^1(\pi \rightarrow \pi^*)$ is comparable to that of C=C double bond:

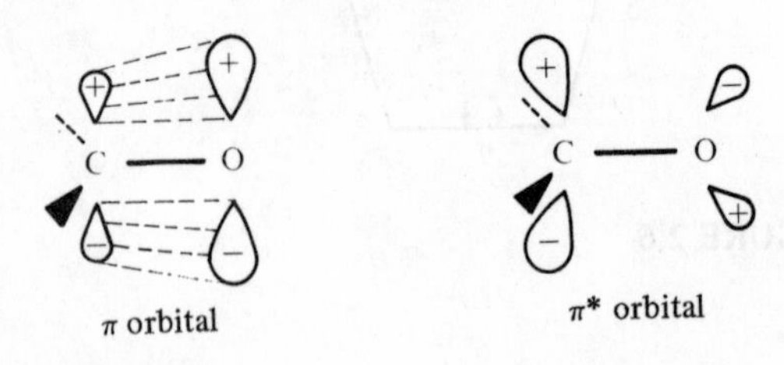

FIGURE 2.8

In the state $^1(n \rightarrow \sigma^*)$ the electron density on the carbon increases, while that on the oxygen decreases, with the result that the carbon acquires a partial negative charge, while the oxygen acquires a partial positive charge.

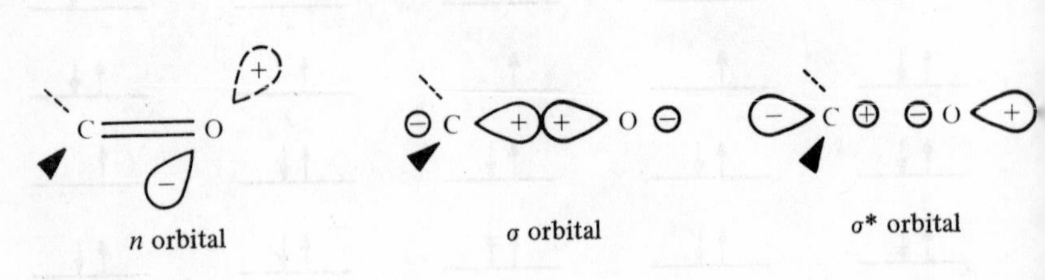

FIGURE 2.9

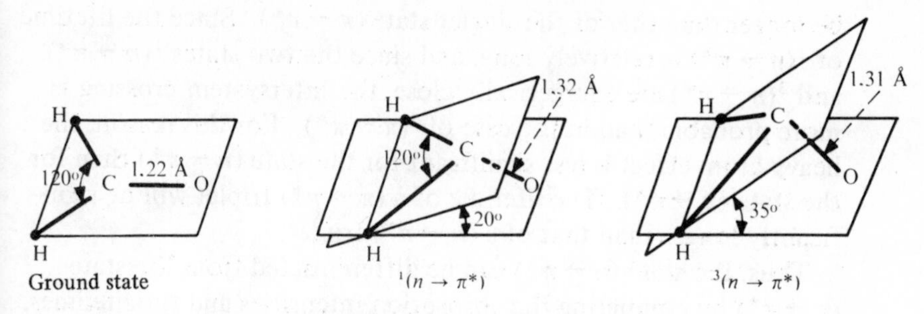

FIGURE 2.10

The state ($n \rightarrow \pi^*$) is of special importance, since it is responsible for most photochemical reactions involving carbonyl groups. It resembles the state ($n \rightarrow \sigma^*$), but obviously requires a smaller transition energy; in this state there is also a partial positive charge on the oxygen atom, and a number of photochemical reactions can be interpreted in terms of this electronic density. Figure 2.10 shows the geometrical differences between the ground state and the excited states $^1(n \rightarrow \pi^*)$ and $^3(n \rightarrow \pi^*)$ of formaldehyde.

3. Experimental differentiation of state ($\pi \rightarrow \pi^*$) from state ($n \rightarrow \pi^*$)

One of the factors of the electronic transition probability P_0 involves the overlapping of the orbitals (Chapter 1). This factor is close to unity for a $\pi \rightarrow \pi^*$ transition, in which the π and π^* orbitals overlap to a considerable extent, but is very small (about 10^{-4}) for a $n \rightarrow \pi^*$ transition, in which the n and the π^* orbitals hardly overlap at all. As a result, the molar extinction coefficient (ϵ_{max}) will be distinctly smaller for a $n \rightarrow \pi^*$ than for a $\pi \rightarrow \pi^*$ transition:

$$(\varepsilon_{max})_{n \rightarrow \pi^*} < 100$$

for carbonyl. Since the lifetime is inversely proportional to the absorption intensity, the lifetime of the singlet state ($n \rightarrow \pi^*$) will be longer than that of the singlet state ($\pi \rightarrow \pi^*$). Since the lifetime of $^1(n \rightarrow \pi^*)$ is relatively long, and since the two states $^1(n \rightarrow \pi^*)$ and $^3(n \rightarrow \pi^*)$ are energetically close, the intersystem crossing is more probable than in the case of $^1(\pi \rightarrow \pi^*)$. For this reason, the heavy atom effect is less significant for the state ($n \rightarrow \pi^*$) than for the state ($\pi \rightarrow \pi^*$). The lifetime of a ($\pi \rightarrow \pi^*$) triplet will be significantly longer than that of a ($n \rightarrow \pi^*$) triplet.

Thus, the state ($n \rightarrow \pi^*$) can be differentiated from the state ($\pi \rightarrow \pi^*$) by comparing the absorption intensities and the lifetimes. Another way of differentiating between the two states is to compare the effects of the solvent on the absorption spectra.

In a state ($n \rightarrow \pi^*$), the unshared electrons of the n orbital can be solvated by a polar solvent: the ground state is thus stabilized and the transition energy increases. Accordingly, the absorption wavelength in a polar solvent will be shorter than in a nonpolar solvent. This is one of the reasons for the "blue shift" of ($n \rightarrow \pi^*$) absorption bands in a polar solvent. Another effect of the polar solvent will be to widen the vibrational fine structure, which is usually sharp in a nonpolar solvent. In a nonpolar solvent the energy difference between two different vibrational levels can be measured and it can be confirmed that such differences are due, say, to a carbonyl group; in a polar solvent the fine structure cannot in general be distinguished.

The fine structure of the state ($\pi \rightarrow \pi^*$) is generally blurred in all solvents, and the effect of solvent polarity is negligible. In polar solvents an effect opposite to that of a $n \rightarrow \pi^*$ transition is observed: there is a "red shift" due to dipole interaction.

A synopsis of the different criteria for distinguishing between the two states is given in Table 2.1.

Another difference between the states ($n \rightarrow \pi^*$) and ($\pi \rightarrow \pi^*$) is the direction of the transition moment. In a singlet–singlet transition, the moment is perpendicular to the plane of the molecule in the state ($n \rightarrow \pi^*$), while it is parallel to the plane of the molecule in the state ($\pi \rightarrow \pi^*$).

Finally, there are significant geometrical differences between the two states: in the state ($\pi \rightarrow \pi^*$), the geometry of the excited

TABLE 2.1. Properties of the states ($n \rightarrow \pi^*$) and ($\pi \rightarrow \pi^*$)

	($n \rightarrow \pi^*$)	($\pi \rightarrow \pi^*$)
Absorption maximum ϵ_{max}	10–10^3 (10–10^2 for C=O)	10^3–10^5
Vibrational structure	Sharp in nonpolar solvents, blurred in polar solvents	Blurred; no solvent effect
Shift of absorption band in polar solvent	Blue	Red, or no shift
Triplet–singlet energy difference	Small	Large
Lifetime of S_1 singlet	10^{-6}–10^{-5} sec	10^{-9}–10^{-7} sec
Lifetime of triplet	$\sim 10^{-3}$ sec	10^{-1}–10 sec
Quantum yield of fluorescence	$\Phi_f < 0.01$	$\Phi_f = 0.05$–0.5
Quantum yield of phosphorescence	$\Phi_p = 0.05$–0.5	$\Phi_p = 0.05$–0.5

state is quite different from that of the ground state (a 90° rotation around the double bond as axis), whereas in the state ($n \rightarrow \pi^*$) the geometry of the excited states is only little different from that of the ground state.

Bibliography

1. Turro, N. J. Molecular Photochemistry. W.A. Benjamin, Inc., N. Y. 1965.
2. Hercules, D. M. Fluorescence and Phosphorescence Analysis. Interscience Publishers, N. Y. 1966.

3. Parker, C. A. Advances in Photochemistry. Edited by W. A. Noyes, Jr., G. S. Hammond, and J. N. Pitts, Jr., 2 : 305–383. Interscience Publishers, N. Y. 1964.
4. McGlynn, S. P., F. J. Smith, and G. Cilento. – Photochemistry and Photobiology, 3 : 269. 1964.
5. Wagner, P. J. and G. S. Hammond. Advances in Photochemistry. Edited by W. A. Noyes, Jr., G. S. Hammond, and J. N. Pitts, Jr., 5 : 21. Interscience Publishers, N. Y. 1968.
6. Coulson, C. A. Reactivity of the Photoexcited Organic Molecules, p.1, Interscience Publishers, N. Y. 1967.
7. Noyes, Jr., W. A. and I. Unger. – Pure Appl. Chem., 9 : 461. 1964.
8. Guilbault, G. G. Fluorescence. M. Dekker, Inc., N. Y. 1967.
9. Bowen, E. J. Luminescence in Chemistry. D. Van Nostrand Company Ltd., London. 1968.
10. Parker, C. A. Photoluminescence of Solutions. Elsevier Publishing Company, Amsterdam. 1968.

Chapter 3

ELECTRONIC ENERGY TRANSFER

I. PROCESSES WHICH COMPETE WITH LUMINESCENCE

We have already seen that luminescence is not the only way in which the excited states of organic molecules can be deactivated. Other processes compete with the luminescence process: photochemical processes, which will be discussed in Chapter 4, and photophysical processes, such as internal conversion and intersystem crossing (see above), as well as the energy transfer now to be discussed.

A process will only be capable of competing with luminescence if its rate constant k_p is higher than that of the quenching of luminescence, i.e., larger than the reciprocal lifetime of the excited state:

$$k_p > 1/\tau .$$

1. Importance of the lowest excited states

Figure 3.1 shows the different electronic states of a molecule which are involved in photophysical and photochemical processes, with the corresponding rate constants. The initial excited state S_n

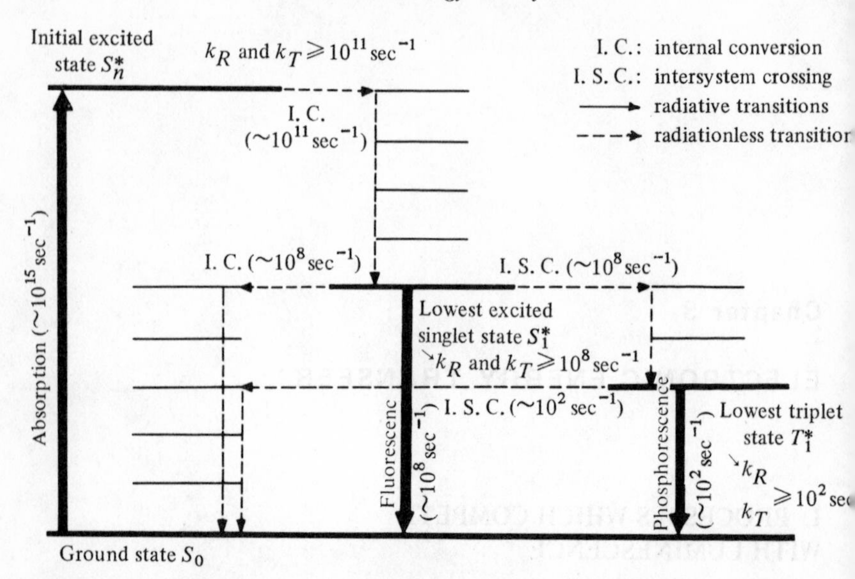

FIGURE 3.1. Schematic representation of electronic transitions of an organic molecule. The orders of magnitude of the rate constants are shown between parentheses. The limiting values of the reaction rate constant k_R and of the energy transfer rate constant k_T are shown for each excited state.

may be a vibrational level of the lowest excited singlet state S_1, or a higher energy singlet state; in all cases, the return to the zero vibrational level of S_1 by internal conversion and vibrational relaxation will last 10^{-13}–10^{-11} seconds. No process can take place at the initial excited state, unless the rate constant of the process is greater than 10^{11} sec^{-1}. The only such effect which has been observed is energy transfer in rigid media. The probability of observing a bimolecular chemical reaction starting from the initial excited state is practically zero: were such a reaction to occur, the molecule could only react with its immediate neighbor, since the diffusion rate constant in the solvent is about 10^9 sec^{-1}. Very rapid monomolecular rearrangements are the only type of reaction which could possibly occur.

We may say, accordingly, that in most cases nothing will happen until the zero vibrational level of the lowest singlet state is attained.

The average lifetime of this lowest singlet state is about 10^{-8} sec, and sometimes even 10^{-6} sec, which means that both energy transfer and chemical reactions become possible. The rate constant of a diffusion-controlled process is of the order of 10^{-9} sec; accordingly, this type of process can be observed starting from the lowest singlet state. Fast reactions, with rate constants higher than 10^8 sec^{-1}, are the only ones which can take place in the singlet state; it follows that this state will be only of minor importance as far as photochemical reactions are concerned.

The lowest triplet state has a long lifetime, of the order of 10^{-2} sec, and sometimes as much as several seconds. Starting from this state, energy transfers and chemical reactions will be the major processes. Thermal deactivation in the medium and intersystem crossing to the ground state will be the only competing processes at ambient temperatures.

2. Electronic origin of the lowest excited state

In the case of a molecule with both $(\pi \rightarrow \pi^*)$ and $(n \rightarrow \pi^*)$ excited states it is very important to know which one is the lower, since it is the identity of this state which will determine the reactivity of the molecule. In a polar solvent the excitation energy of singlet $(n \rightarrow \pi^*)$ states increases, while that of $(\pi \rightarrow \pi^*)$ states remains unchanged or decreases slightly. Thus, certain molecules, such as quinoline or acridine, have* $^1(n \rightarrow \pi^*)$ as their lowest singlet state in a non-polar solvent, and $^1(\pi \rightarrow \pi^*)$ in a polar solvent: accordingly, fluorescence will be very feeble in a non-polar solvent, in which the intersystem crossing is easy, but will be marked in a polar solvent.

It is also important to note that internal conversions* can occur even between different singlet states: thus, for instance, there may be internal conversion between the initial excited state $^1(\pi \rightarrow \pi^*)$ and the lowest excited state $^1(n \rightarrow \pi^*)$; the same may well apply to

* A number of workers speak of "intramolecular energy transfer" to denote internal conversion or intersystem crossing between different electronic states.

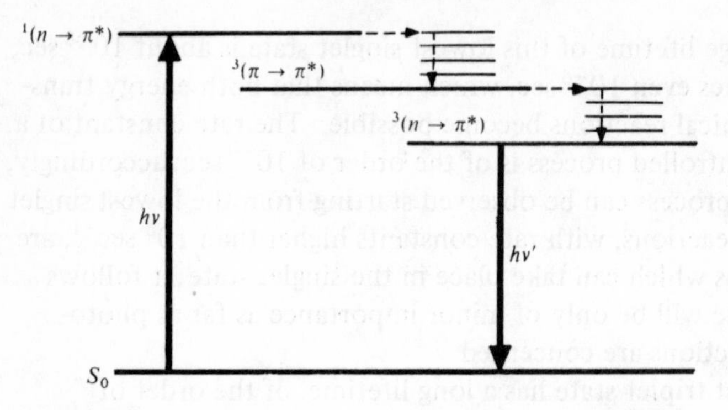

FIGURE 3.2. Aromatic carbonyl compounds.

triplet states. It was calculated by El Sayed that intersystem crossing* between two states of different electronic origins is distinctly favored; thus, the crossing

$$^1(\pi \to \pi^*) \rightsquigarrow {}^3(n \to \pi^*)$$

is 1000 times faster than the crossing

$$^1(\pi \to \pi^*) \rightsquigarrow {}^3(\pi \to \pi^*)\,.$$

In cases in which this rule is not exactly obeyed, intersystem crossings proceed more readily in molecules having both $(n \to \pi^*)$ and $(\pi \to \pi^*)$ states than in those which have $(\pi \to \pi^*)$ only. This is why most aromatic carbonyl compounds have very large (10^{10} sec^{-1}) intersystem crossing rate constants, and emit an intense phosphorescence and no fluorescence: in these compounds, the triplet state $^3(\pi \to \pi^*)$ has an energy which is intermediate between $^1(n \to \pi^*)$ and $^3(n \to \pi^*)$ (Figure 3.2). The intersystem crossing

$$^1(n \to \pi^*) \rightsquigarrow {}^3(\pi \to \pi^*)$$

* A number of workers speak of "intramolecular energy transfer" to denote internal conversion or intersystem crossing between different electronic states.

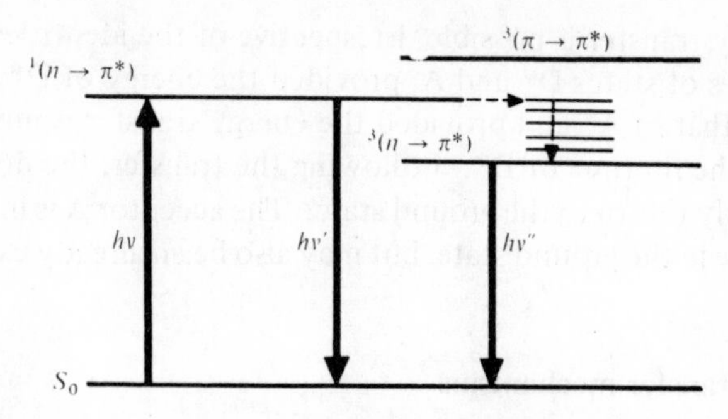

FIGURE 3.3. Aliphatic carbonyl compounds.

will be very rapid, and the internal conversion to $^3(n \rightarrow \pi^*)$ will follow within 10^{-12} sec. In aliphatic carbonyl compounds, the triplet state $^3(\pi \rightarrow \pi^*)$ will generally have a larger energy than state $^1(n \rightarrow \pi^*)$ and the intersystem crossing will take place directly (Figure 3.3):

$$^1(n \rightarrow \pi^*) \rightsquigarrow {}^3(n \rightarrow \pi^*) .$$

The rate constant of the intersystem crossing will be only 10^7 sec^{-1} and fluorescence may be observed.

II. INTERMOLECULAR ELECTRONIC ENERGY TRANSFER

The electronic energy transfer between an excited "donor" molecule D* and an acceptor molecule A is a bimolecular process of deactivation of the donor. The deactivation of D* by A is known as quenching, while the excitation of A by D* is known as photosensitization, but the process

$$D + h\nu \rightarrow D^*$$
$$D^* + A \rightarrow A^* + D$$

involved is the same.

An energy transfer is possible, irrespective of the identities or multiplicities of states D* and A, provided the energy of D* is larger than that of A* and provided the energy transfer is more rapid than the lifetime of D*. Following the transfer, the donor will invariably return to the ground state. The acceptor A is in general a molecule in the ground state, but may also be an already excited molecule.

1. Energy transfer mechanisms

Many types of energy transfer exist, depending on the distance between the donor and the acceptor molecule; moreover, certain mechanisms are specific to the liquid phase, while others are observed for transfers in the solid phase.

A) *Energy transfers in the liquid phase*

A.1) *Radiative Transfer.* This is the simplest case. If the donor D* emits luminescence radiation of wavelength corresponding to the absorption of the acceptor A, the energy transfer may take place by the reabsorption of the luminescence:

$$D + h\nu \rightarrow D^*$$

$$D^* \rightarrow D + h\nu'$$

$$A + h\nu' \rightarrow A^* .$$

At long distances (> 100 Å) this is the only conceivable type of energy transfer. Its rate is not affected by the viscosity of the solvent, and the lifetime of the excited state of the donor is also unchanged by the energy transfer.

A.2) *Radiationless Energy Transfers.* Two cases should be considered:

If the molecules are not in contact with each other, the transfer will be of the long-distance dipole–dipole type (50–100 Å).

If the molecules are in contact with each other, energy transfer will take place by the collision mechanism, either through overlapping of the orbitals of two molecules, or through an intermediate charge transfer complex. The range of this transfer type is 10–15 Å and the rate of the process is obviously determined by the diffusion rate of the molecules in the solvent.

a) *Dipole–dipole energy transfer.* The electron systems of D* and A can be treated as coupled mechanical oscillators (i.e., oscillators vibrating at the same frequency); if one of the oscillators is excited, the excitation energy is shared with the other oscillator. In the same manner, if D* and A have the same vibration frequency, the energy may be transferred by resonance, since in quantum mechanics the degenerate systems

$$D^* \text{ --- } A \leftrightarrow D \text{ --- } A^*$$

are identical. If the lowest vibrational state of A* is lower than the lowest vibrational state of D*, A* is internally converted to its lowest vibrational level at a very rapid rate, and energy transfer will only be possible from D* to A (Figure 3.4).

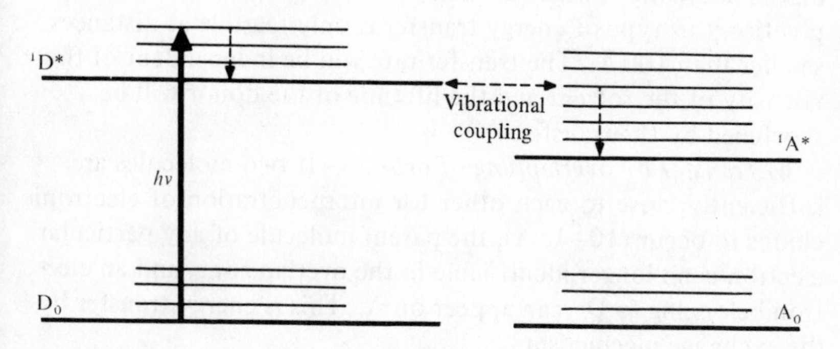

FIGURE 3.4

Just as in radiative transfer, this transfer can only take place if the emission spectrum of D* overlaps with the absorption spectrum of A.

The transfer will be favored if the emission by D* and the absorption by A are allowed transitions (singlet–singlet or triplet–triplet). If the transition of D* is forbidden while that of A is permitted, the transfer can still be realized, since the slower rate of transfer will be offset by the longer lifetime of state D*; on the other hand, if the transition of D* is permitted, but that of A is forbidden, the transfer will be very improbable owing to the coincidence of a slow transfer rate with a short lifetime of D*. The singlet–singlet energy transfer

$$^{1}D^{*} + A_{0} \rightarrow {}^{1}A^{*} + D_{0}$$

by this mechanism is possible, but not the triplet–triplet transfer

$$^{3}D^{*} + A_{0} \rightarrow {}^{3}A^{*} + D_{0}\,.$$

Since this transfer mechanism is due to the coupling of transition dipole moments of D* and A, its probability will be inversely proportional to R^6 (where R is the distance between D* and A). This probability is directly proportional to the square of dipole–dipole interaction, which is itself inversely proportional to R^3. In practice, this type of energy transfer is only feasible at distances smaller than 100 Å. The transfer rate will be independent of the viscosity of the solvent and the lifetime of the donor will be shortened by the transfer.

b) *Transfer by overlapping of orbitals.* If two molecules are sufficiently close to each other for interpenetration of electronic clouds to occur (10–15 Å), the parent molecule of any particular electron is no longer identifiable in the overlap zone, and an electron belonging to D* can appear on A. This is energy transfer by the exchange mechanism.

Unlike the resonance mechanism (see a) above), this mechanism allows an energy transfer involving a transition which is forbidden for D* and for A, and the very important triplet–triplet energy transfer

$$^{3}D^{*} + A_{0} \rightarrow {}^{3}A^{*} + D_{0}$$

will take place. Singlet–singlet energy transfers can also take place by this mechanism.

This process is controlled by the diffusion rate of the molecules in the solvent; the lifetime of D* must be longer than 10^{-9} sec. An increase in the viscosity of the solvent will reduce the transfer rate. The lifetime of the donor D* will decrease as a result of the transfer.

c) *Energy transfer by way of an intermediate charge transfer complex.* Charge transfer complexes will be studied in more detail in Chapter 13, but their definition will be given here, since they are relevant to electronic energy transfers. According to Mulliken, a charge transfer complex is a molecular complex formed between an electron donor D and an electron acceptor A whose electronic spectrum contains an absorption band which is not present either in the spectrum of A alone or in the spectrum of D alone. This band is known as a "charge transfer transition," since it may be assumed, as a first approximation, that it originates from an electron transfer from D to A:

$$(D + A) \xrightarrow{h\nu} (D^+ \text{ --- } A^-)^* .$$

However, an electron transfer of this kind need not necessarily be accompanied by the appearance of an absorption band. Molecules in the ground state generally have little effect on one another and their complexing only becomes significant if one of them is in the excited state. The most reasonable explanation is that the transfer takes place during a collision. We then have the following possible energy transfer mechanism:

$$D^* + A \rightarrow (D^+ \text{ --- } A^-)^*$$

$$(D^+ \text{ --- } A^-)^* \rightarrow D + A^* .$$

This type of energy transfer plays an important part in biochemistry (Chapter 13), and in the preparation of excited oxygen (Chapter 8).

B) *Solid phase energy transfer*

The radiative mechanism and the dipole–dipole coupling mechanism may be operative in the solid phase. The following two other mechanisms typically occur in the solid phase.

B.1) *Exciton Theory.* This is a radiationless, noncollisional process of energy transfer, of which dipole–dipole transfer is a particular case. The very rapid energy transfer in crystals from one molecule to another can be explained by migration of an excitation wave ("exciton").

B.2) *Theory of Semiconductors.* Electronic energy can be transmitted in the crystal by the migration of electrons and of positive holes in the crystal. This effect is very often observed in mineral crystals, and is probably very important in organic chemistry as well.

2. Energy transfers utilized in organic photochemistry

Only energy transfers in liquid solutions will be considered here. The following types of energy transfer are of importance to the photochemist:

A) singlet–singlet transfer: $^1D^* + {}^1A_0 \rightarrow {}^1A^* + {}^1D_0$
B) triplet–singlet transfer: $^3D^* + {}^1A_0 \rightarrow {}^1A^* + {}^1D_0$
C) triplet–triplet transfer: $^3D + {}^1A_0 \rightarrow {}^3A^* + {}^1D_0$

Triplet annihilation $^3D^* + {}^3D^* \rightarrow {}^1D^* + {}^1D_0$ is an energy transfer which is responsible for delayed type P fluorescence (Chapter 2). Another very important energy transfer corresponds to sensitized excitation of molecular oxygen and will be discussed in Chapter 8. It may be represented by the following equation, in which $^3A_0$ is the triplet ground state of oxygen:

36

$$^3D^* + {}^3A_0 \rightarrow {}^1A^* + {}^1D_0 .$$

Except for this transfer and transfers of type B) above, all energy transfers obey Wigner's rule of conservation of spin, since the resultant spin of the system remains unchanged.

A) *Singlet–singlet transfer*

$$^1D^* + {}^1A_0 \rightarrow {}^1A^* + {}^1D_0 .$$

This type of transfer may take place both by resonance and by exchange mechanisms. Since the lifetime of the singlet is very short, this type of energy transfer will not be favored; in any case, since the excited singlet state can be easily produced by absorption, its production by photosensitization is of no practical interest. However, this type of transfer may be operative in inhibiting the excited singlet state; in particular, it may be the explanation for the inhibition of fluorescence of certain molecules by the impurities they contain.

$$^1D_0 + h\nu \rightarrow {}^1D^*$$
$$^1D^* \nrightarrow D + h\nu' ,$$
$$^1D^* + {}^1A_0 \rightarrow {}^1A^* + {}^1D_0 .$$

If the inhibitor A fluoresces, the fluorescence observed will be that of A and not of D; if it does not, no fluorescence will be observed.

B) *Triplet–singlet transfer*

$$^3D^* + {}^1A_0 \rightarrow {}^1A^* + {}^1D_0 .$$

This kind of transfer is usually forbidden by Wigner's rule, but we have seen that it can be produced by the resonance mechanism. It must be borne in mind that if other modes of deactivation of the energy of the donor $^3D^*$ are also forbidden, a triplet–singlet transfer may become significant: if $^3D^*$ does not enter a photochemical reaction and if a triplet–triplet transfer does not take place, $^3D^*$ can only lose its activation energy by phosphorescence and by intersystem crossing to the ground state, which are also forbidden processes.

A transfer of this kind could be identified between phenanthrene P (donor) and fluorescein F (acceptor). It was observed that the duration of phosphorescence emitted by phenanthrene was shortened; fluorescence emitted by fluorescein appeared instead, its lifetime being of the same order as the phosphorescence of phenanthrene:

$P + h\nu \rightarrow {}^1P^*$	absorption
${}^1P^* \rightsquigarrow {}^3P^*$	intersystem crossing
${}^3P^* \rightarrow P + h\nu'$	phosphorescence of phenanthrene
${}^3P^* + F \rightarrow {}^1F^* + P$	triplet–singlet energy transfer
${}^1F^* \rightarrow F + h\nu''$	fluorescence of fluorescein

The rate constant of this energy transfer is independent of the viscosity of the solvent; thus, the process is not diffusion-controlled, but proceeds by a resonance mechanism.

C) *Triplet–triplet energy transfer*

$$ {}^3D^* + A_0 \rightarrow {}^3A^* + D_0 . $$

This transfer will be efficient only if the energy of ${}^3D^*$ is larger than that of ${}^3A^*$. This type of transfer was discovered by Ermolaev and Terenin, who observed the sensitized phosphorescence of certain aromatic compounds. It was applied by Hammond to photosensitized reactions. This type of transfer proceeds by an exchange mechanism only. This is the most important kind of energy transfer in organic photochemistry, for the following reasons:

it is favored by the long lifetime of the triplet donor;

the overlapping of the emission spectrum of the donor with the absorption spectrum of the acceptor need not be as complete as for a singlet–singlet transfer;

if the donor is appropriately chosen, it is possible to study the triplet state of a molecule, especially its chemical reactivity, without interference from its singlet state.

The rates of energy transfer can be related to the energy difference between the triplets of the donor and of the acceptor, as follows:

The effectiveness of an acceptor or an inhibitor does not depend on its molecular structure, but on the energy of its lowest triplet.

TABLE 3.1

Compound	E_{T_1}, kcal · mole^{-1}
Benzene	85
Toluene	83
Phenol	82
Benzoic acid	78
Benzonitrile. Aniline	77
Propiophenone	75
Acetophenone. Xanthone. Diisopropyl ketone. Diphenyl sulfide. *p*-Dichlorobenzene	74
Isobutyrophenone	73
Benzaldehyde. Diphenylselenium. 1,3-Diphenyl-2-propanone. Diphenylamine	72
Triphenylmethyl phenyl ketone	71
Carbazole. Triphenylamine. Hexachlorobenzene. Diphenylene oxide. Dibenzothiophene.	70
Benzophenone. Thiophene. 4,4-Diphenylcyclohexadienone. 1,2-Dibenzoylbenzene	69
p-Diacetylbenzene. 4,4′-Dichlorobenzophenone. Fluorene	68
Triphenylene. 9-Benzoylfluorene	67
p-Cyanobenzophenone	66
Biphenyl. Thioxanthone	65
Phenylglyoxal	63
Phenanthrene. Anthraquinone. Flavone. Quinoline. *cis*-Stilbene. 1-Naphthoflavone. Ethyl phenyl glyoxalate	62
Naphthalene. 4-Acetylbiphenyl. 4,4′-*bis*-(Dimethylamino)-benzophenone	61
Nitrobenzene. 2-Naphthyl phenyl ketone. 2-Naphthaldehyde. Methyl 1-naphthyl ether	60
2-Acetonaphthone. 1-Bromonaphthalene. 1-Naphthol	59
1-Naphthylamine. Acridine Yellow. *m*-Nitroacetophenone	58
1-Naphthyl phenyl ketone. Chrysene	57
1-Acetonaphthone. 1-Naphthaldehyde. 5,12-Naphthacenequinone	56
Biacetyl. Acetylpropionyl. Coronene. 1-Nitronaphthalene. *p*-Nitroaniline	55

TABLE 3.1 (continued)

Compound	E_{T_1}, kcal · mole^{-1}
Benzil	54
Fluorenone	53
1,2,5,6-Dibenzanthracene	52
Fluorescein (acid)	51
trans-4-Nitrostilbene	50
Pyrene	49
Pentaphene	48
1,2-Benzanthracene	47
11,12-Trimethylenetetraphene. 1,12-Benzperylene	46
Phenazine	44
Eosin. Iron dicyclopentadienyl	43
Anthracene. 3,4-Benzpyrene	42
Thiobenzophenone	40
Crystal Violet	39
Naphthacene	29

Studies of inhibition and of sensitization make it possible to identify the excited states involved and the reaction mechanisms.

The collisional triplet–triplet transfer makes asymmetric induction possible. If the triplet–triplet transfer proceeds by a collision mechanism, it can be affected by stereochemical factors. According to Hammond, an optically active donor could discriminate between the *d*- and *l*-isomers of a racemic acceptor. He demonstrated the presence of such an asymmetric induction in the case of photochemical *cis-trans* isomerization of *d, l-trans*-1,2-diphenylcyclopropane by an optically active sensitizer: the *trans*-isomer which had been isolated from the irradiated *cis-trans* mixture displayed marked optical activity. Hammond also gave other examples of asymmetric induction of this kind. According to him, a non-classical transfer with a non-vertical transition (*vide infra*) is more probable, since its rate is much slower than the limiting rate of diffusion.

It will be possible to populate the triplet state of molecules which give very low quantum yields in intersystem crossing.

The same molecule can act as triplet donor with respect to molecules with a lower triplet energy, and as triplet acceptor with respect to molecules with a higher triplet energy. Table 3.1 lists the lowest triplet energies for a number of molecules utilized in triplet–triplet transfers.

3. Energy transfers in organic photochemistry

A) *Photosensitized reactions*

Photosensitized reactions usually involve a triplet–triplet rather than a singlet–singlet energy transfer. The choice of the donor (or sensitizer) is very important for the following reasons:

in order to have a high triplet quantum yield, the intersystem crossing of the sensitizer must be effective;

the sensitizer must be chemically inert;

the energies of the lowest excited triplet and singlet states of the sensitizer must be known, and the choice of the sensitizer will depend on the acceptor studied; the donor triplet must have a higher energy than the acceptor triplet for the transfer to be possible but, on the other hand, the first excited singlet of the acceptor must (or at least should) have a higher energy level than the sensitizer in order to prevent direct absorption of light by the acceptor (Figure 3.5).

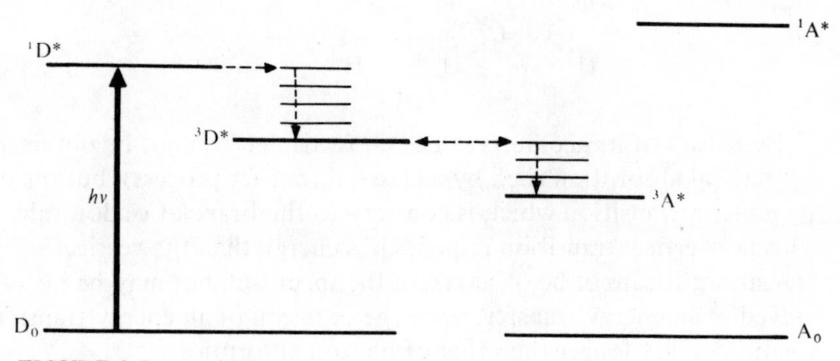

FIGURE 3.5

The rule of relative energies of donor and acceptor triplets ($E_{3D^*} > E_{3A^*}$) is always obeyed, despite occasional appearances to the contrary. Thus, photosensitized reactions in which the lowest acceptor triplet has a higher energy level than the lowest donor triplet have been reported, but such anomalies may have two explanations:

a) the energy of the acceptor triplet involved in the transfer is lower than that of its lowest triplet (theory of "ghost triplet");

b) the triplet of the donor involved in the transfer is not its lowest triplet but a higher energy state T_2.

a) *Theory of "ghost triplet."* This theory was put forward by Hammond to explain the photosensitized *cis-trans*-isomerization of stilbene by sensitizers whose energy was too low. Sensitizers with $E_T < 57$ kcal · mole^{-1}, which is the lowest triplet of *cis*-stilbene, would not be expected to convert *cis*-stilbene to the *trans*-isomer, but this conversion does actually occur. Hammond assumes a "ghost triplet" state of stilbenes, the energy of which is smaller than that of the *cis*-triplet (57 kcal) and the *trans*-triplet (46 kcal), and the geometry of which is intermediate between the two. If triplet energies are plotted as a function of the dihedral angle of C—Φ bonds, the curve will have a minimum* corresponding to the "ghost triplet" at a value of the dihedral angle of about 90°:

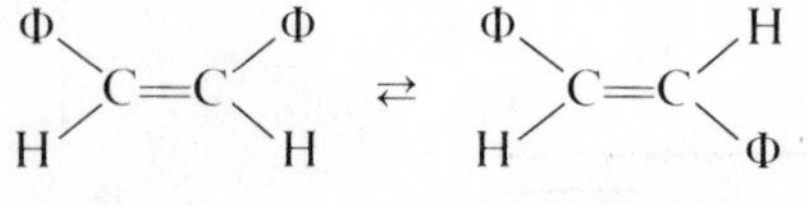

$$\begin{matrix}\Phi & & \Phi \\ & C{=}C & \\ H & & H\end{matrix} \rightleftarrows \begin{matrix}\Phi & & H \\ & C{=}C & \\ H & & \Phi\end{matrix}$$

By reason of its geometry, this "ghost triplet" cannot be obtained by vertical absorption (i.e., by a classical transfer process), but only by a direct transition which is contrary to the Franck-Condon rule. This nonvertical transition requires less energy than the vertical transition: it cannot be observed in the spectrum, but may be involved in an energy transfer, since the duration of an energy transfer is considerably longer than that of photon absorption.

* See *cis-trans*-isomerization, Chapter 11, Figure 11.3.

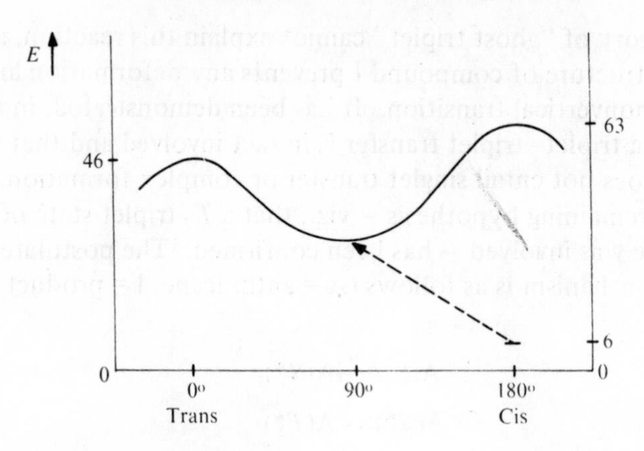

FIGURE 3.6

Hammond also experimentally found that the nonvertical transition

cis-stilbene → ghost triplet

requires less energy than the vertical transition

trans-stilbene → ghost triplet

The reason for it is steric assistance of the eclipsed phenyls in the *cis*-isomers (6 kcal · mole^{-1}).

Unlike the classical triplet–triplet energy transfer, in the case of stilbenes the transfer is not diffusion-controlled.

b) *Participation of the T_2 state.* Photochemical isomerization of compound I has been studied. Even though the triplet energy of I (59–68 kcal · mole^{-1}) is larger than that of the lowest triplet of anthracene (E_T = 42.5 kcal · mole^{-1}), its isomerization can be sensitized by anthracene /3/:

CF$_3$ F$_3$C I $\xrightarrow[\text{sens.}]{h\nu}$ CF$_3$ F$_3$C + CF$_3$ CF$_3$ + CF$_3$ CF$_3$

The theory of "ghost triplet" cannot explain this reaction, since the rigid structure of compound I prevents any deformation and forbids a nonvertical transition. It has been demonstrated, in any case, that a triplet–triplet transfer is in fact involved and that the reaction does not entail singlet transfer or complex formation. The only remaining hypothesis – viz., that a T_2 triplet state of anthracene was involved – has been confirmed. The postulated reaction mechanism is as follows (A = anthracene, I = product I):

$$\mathrm{A}(S_0) \xrightarrow{h\nu} \mathrm{A}(S_1^*)$$

$$\mathrm{A}(S_1^*) \rightsquigarrow \mathrm{A}(T_2^*)$$

$$\mathrm{A}(T_2^*) \rightsquigarrow \mathrm{A}(T_1^*) \rightsquigarrow \mathrm{A}(S_0)$$

$$\mathrm{A}(T_2^*) + \mathrm{I}(S_0) \rightarrow \mathrm{A}(S_0) + \mathrm{I}(T_1^*)$$

$$\mathrm{I}(T_1^*) \rightarrow \text{isomers of I.}$$

The relative energies of the different excited states of anthracene (kcal · mole^{-1}) are shown in Figure 3.7.

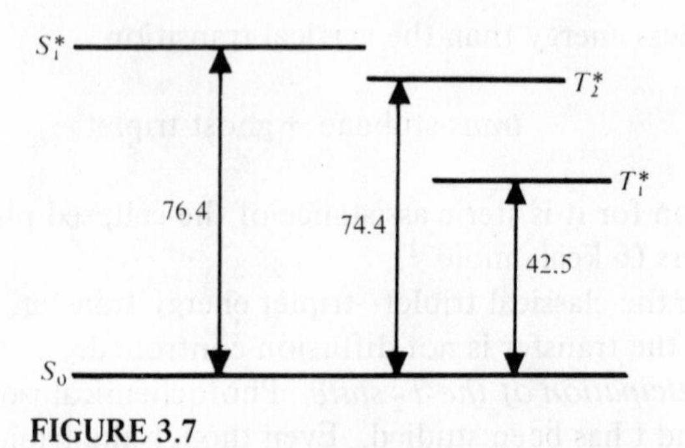

FIGURE 3.7

The lifetime of the T_2 state of 9,10-dibromoanthracene was calculated (Stern-Volmer equation, see below) as

$$\tau_{T_2} = 2 \times 10^{-11} \text{ sec.}$$

B) *Inhibition of reactions. Stern-Volmer equation*

The study of the inhibition (quenching) of a reaction is a powerful tool in determining its mechanism, and particularly in identifying the excited states involved. The selection of a satisfactory inhibitor to some extent resembles the choice of a satisfactory sensitizer:

it should be chemically inert. Olefins, which are very often used as "inhibitors," are not inert to free radicals;

as regards the relative energies of donor and acceptor, considerations which are valid for sensitized reactions are also valid for inhibited reactions (here A is the inhibitor and D is the reactant) (Figure 3.5).

The kinetic study of the inhibition of a photochemical reaction (or of luminescence) makes it possible to determine the lifetime of an excited state by using the Stern-Volmer equation. The mechanism of quenching of the luminescence emitted by an excited molecule D* by an inhibitor A can be written as follows:

			Rate
(1)	$D + h\nu \longrightarrow D^*$	(absorption)	I_a
(2)	$D^* \xrightarrow{k_1} D + h\nu'$	(emission)	$k_1 . [D^*]$
(3)	$D^* + A \xrightarrow{k_2} A^* + D$	(transfer)	$k_2 . [D^*] . [A]$
(4)	$D^* \xrightarrow{k_3} D$	(internal conversion)	$k_3 . [D^*]$

$$\frac{d[D^*]}{dt} = I_a - k_1.[D^*] - k_2.[D^*].[A] - k_3.[D^*]$$

$$= I_a - [D^*].(k_1 + k_3 + k_2.[A]) .$$

If the concentration of D* becomes stationary, we have $\frac{d[D^*]}{dt} = 0$, i.e.,

$$I_a = [D^*].(k_1 + k_3 + k_2.[A]). \tag{3.1}$$

According to (1.4), the quantum yield of emission by D* is

$$\Phi = \frac{k_1.[D^*]}{I_a}$$

or

$$\Phi = \frac{k_1}{k_1 + k_3 + k_2.[A]}. \tag{3.2}$$

In the absence of the inhibitor [A] = 0, so that

$$\Phi_0 = \frac{k_1}{k_1 + k_3}. \tag{3.3}$$

The Stern-Volmer equation is obtained by dividing Φ_0 by Φ:

$$\frac{\Phi_0}{\Phi} = \frac{k_1 + k_3 + k_2.[A]}{k_1 + k_3} = 1 + \frac{k_2.[A]}{k_1 + k_3}. \tag{3.4}$$

Putting $\tau = \frac{1}{k_1 + k_3}$, we obtain the final form of the Stern-Volmer equation:

$$\frac{\Phi_0}{\Phi} = 1 + k_2.\tau.[A], \tag{3.5}$$

where τ is the experimental lifetime of D* in the absence of A.

The plot of Φ_0/Φ as a function of [A] will be a straight line with the slope $k_2\tau$. If the lifetime τ of emission by D* in the absence of A is known, k_2 can be experimentally found.

If the lifetime of the excited state D* cannot be directly measured (luminescence), it can be calculated by using the Stern-Volmer equation if the value of k_2 is known. Thus, for instance, the value of k_2 of a diffusion-controlled process is given by

$$k_2 = \frac{8\,RT}{3\,000\,\eta}, \tag{3.6}$$

where η is the solvent viscosity in poises. For diffusion-controlled processes in organic solvents at room temperature, the rate constant will be $10^9 - 10^{10}\,\text{sec}^{-1}$.

III. INTRAMOLECULAR TRANSFER OF ELECTRONIC ENERGY

If a molecule has the excited states $(n \rightarrow \pi^*)$ and $(\pi \rightarrow \pi^*)$, a transition to the lowest excited state is always observed, whatever its electronic origin; the intramolecular energy transfer takes place in the same chromophore or in two conjugated chromophores.

Energy transfer between two non-conjugated chromophores can be considered as a particular case of intermolecular energy transfer in which the two molecules remain at a constant distance from each other, thus facilitating the transfer. Singlet–singlet and triplet–triplet intramolecular transfers have been observed. If a molecule such as

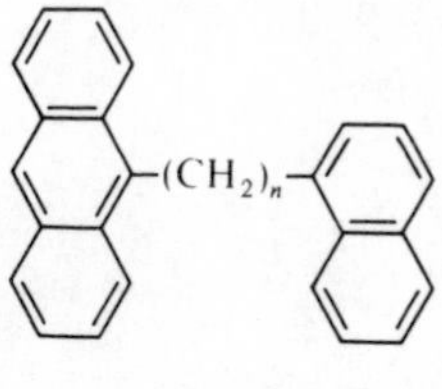

$-(CH_2)_n-$

is excited by 2800 Å radiation, anthracene alone fluoresces, while the naphthalene absorbs most of the radiation; an equimolecular mixture of anthracene and naphthalene, on the other hand, emits both kinds of fluorescence. This example shows that intramolecular energy transfers are much more rapid than inter-molecular transfers.

Molecules such as

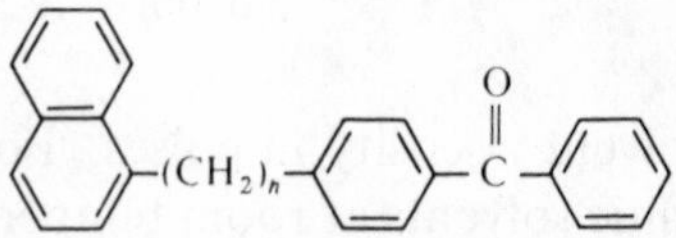

give a triplet–triplet transfer: excitation of the carbonyl by 3660 Å radiation results in a phosphorescence due to naphthalene.

Bibliography

1. El Sayed, M. A. – Accounts of Chemical Research, 1:8. 1968.
2. Wagner, P. J. and G. S. Hammond. Advances in Photochemistry. Edited by W. A. Noyes, Jr., G. S. Hammond, and J. N. Pitts, Jr., **5**: 21. Interscience Publishers, N. Y. 1968.
3. Liu, R. S. H. and J. R. Edman. – J. Am. Chem. Soc., **90**: 213. 1968.
4. Hammond, G. S. Reactivity of the Photoexcited Organic Molecules, pp.119–144. Interscience Publishers, N. Y. 1967.

Chapter 4

PHOTOCHEMICAL REACTIONS

I. INTRODUCTION

To the chemist, photochemical reactions are the most important mode of deactivation of excited states. In addition, the knowledge of photophysical processes of activation and deactivation is indispensable in understanding the mechanism of these reactions. The necessary condition for an excited state to take part in a chemical reaction is that the reaction should be faster than the lifetime of the excited state. Since the lifetimes of triplet states are long, they will give chemical reactions more readily than singlet states.

The physical properties (geometry, electric properties, magnetic properties, thermodynamical properties) of the excited states of any given molecule will in general be different from those of its ground state, and for this reason their chemical reactivities will also be different. Light absorption is a specific process which affects only a part of the molecule; for this reason photochemical reactions may be highly selective. Their selectivity can even be further increased by photosensitization (Chapter 3).

1. "Pseudo" photochemical reactions

The internal conversion $S_1 \rightsquigarrow S_0$, or intersystem crossing $T_1 \rightsquigarrow S_0$ initially consist in a transition to an isoenergetic level of the ground state; since in general the energy of S_1 or T_1 is high, the ground state level thus arrived at contains a very large excess of vibrational energy. The dissipation of this energy by vibrational relaxation is a very rapid process, especially so in solution. Nevertheless, chemical reactions starting from this "hot" ground state are possible; in solution, only very fast reactions (monomolecular reactions and very few bimolecular reactions) can take place, but in vapor phase, especially under low pressures, chemical reactions will be more favored.

These reactions are not photochemical reactions, but thermal reactions, and the required energy could just as well be supplied to the molecule by thermal activation. Secondary reactions taking place by thermal activation will generally be more marked; in fact, thermal activation of a molecule or of a part of it can only be the result of an increase in the total energy of the system. It is generally difficult to show that a chemical reaction proceeds by a mechanism of this type.

2. Primary photochemical processes

An excited organic molecule (singlet or triplet) may undergo a number of different primary photochemical processes, which include:

dissociation to free radicals;
decomposition into molecules;
intramolecular rearrangement;
intramolecular or intermolecular cleavage of a hydrogen atom;
photochemical addition;
photochemical dimerization;
photoionization.

Most of these primary processes will be illustrated in Chapter 7, which deals with the photochemistry of carbonyl compounds.

3. Secondary processes

Very often it is the secondary reactions which yield the final products of photochemical reactions. This happens, in particular, whenever the primary process consists in a dissociation into free radicals. In general, free radicals contain excess energy (electronic energy, since the radical is excited; vibrational energy of the "hot" state; translation energy). In solution, recombination is usually the major reaction, since the radicals formed remain in a solvent "cage," and the rate of recombination is often superior to that of diffusion into the solvent. Thus, in solution, intramolecular processes are favored; an example is the detachment of a hydrogen atom from the molecule by the radical. Intermolecular reactions usually involve solvent molecules. Free alcoholate radicals (photochemistry of organic nitrites, see below) undergo a variety of photochemical reactions.

Chain reactions. The free radicals formed by the primary processes may initiate chain reactions. A chain reaction includes the initiation stage (formation of a primary radical by, say, a photochemical reaction), a propagation stage and a chain termination stage. The number of molecules of the product formed per quantum initially absorbed may be very considerable ($\Phi \gg 1$). A classical example is the chlorination of a hydrocarbon:

initiation $$Cl_2 \xrightarrow{h\nu} 2\,Cl\cdot$$

propagation $$Cl\cdot + RH \longrightarrow R\cdot + ClH$$

$$R\cdot + Cl_2 \longrightarrow RCl + Cl\cdot$$

---

termination $$Cl\cdot + Cl\cdot \longrightarrow Cl_2$$

$$R\cdot + R\cdot \longrightarrow R{-}R$$

$$Cl\cdot + R\cdot \longrightarrow R{-}Cl\,.$$

II. DETERMINATION OF THE MECHANISM OF A PHOTOCHEMICAL REACTION

The determination of the nature and of the yields of primary photophysical or photochemical processes is very important, since in this way a correlation can be made between the spectroscopic properties, the molecular structure and the photochemical reactivity. However, such work is very difficult; various techniques have been proposed and will now be briefly reviewed.* It is usually easier to determine the nature of the secondary reactions, and such determinations furnish valuable information on photochemical mechanisms.

1. Spectroscopic study of reactants

A spectroscopic study is indispensable; the parameters studied are the wavelength, intensity and fine structure of absorption or emission bands. The absorption spectrum will generally provide information on the nature of the initial excitation. Spectroscopic determination of lifetimes of the excited states is very often utilized for the purpose. Quantitative studies of emission spectra provide data on the lifetimes of the excited states, the rates of intersystem crossing, the number and the nature of the excited states, etc. Other techniques employed include fluorescence quenching, and sensitization of fluorescence and of phosphorescence.

2. Purification of reactants

Purification of the reactants employed is highly important. Traces of impurities in the reactants, or traces of oxygen in solutions may significantly alter the quantum yield of a reaction. Degassing of solutions will be discussed in Chapter 14, which deals with the experimental techniques.

3. Identification of reaction products

In studying the mechanism of a reaction, it is important to ensure that the degree of conversion of the reactants does not

* For a more advanced account, the reader is referred to /1/ in Chapter 1, and to /1/ in Chapter 2.

exceed a few percent (in many cases less than one percent), in order to avoid the complications due to the subsequent participation of the reaction products formed in other thermal or photochemical reactions. The techniques used to perform the analysis of the reaction products must therefore be very sensitive, and vapor phase chromatography and spectroscopic methods are most often employed in practice. It will also be obviously necessary to conduct independent experiments at higher degrees of conversion aimed at the isolation of the reaction products with a view to their identification, structure determination, etc., by chemical and physicochemical methods.

4. Determination of quantum yields

This is probably the most useful technique in the study of reaction mechanisms, even though the determination of primary quantum yields is generally difficult. A study of the quantum yield of a given reaction product as a function of the experimental conditions will often determine whether or not the compound is a primary reaction product: if its quantum yield remains unchanged when the experimental conditions (e.g., temperature) are varied, the compound is a primary reaction product.

5. Direct spectroscopic identification of primary products

The chemical species formed by primary processes often have very short lifetimes. Various spectroscopic techniques may be used to identify these products.

A) *Flash photolysis*

A very strong flash of light may result in the formation of a significant concentration of intermediate products, the identity and the secondary reactions of which can be studied by classical spectroscopy.

B) *Photolysis in rigid media*

Intermediate species may be preserved in vitrified solvents at a low temperature or in glasses, e.g., in boric acid, at ordinary temperatures. In particular, photoionization reactions could be observed in a boric acid glass; the glass captures the electron and thus stabilizes the cation formed, which would otherwise undergo very fast recombination.

C) *Mass spectroscopy*

Identification of free radicals formed by a primary process can be effected with the aid of mass spectroscopy, but the technical difficulties are very great.

D) *Infrared spectroscopy*

If a multiple reflection cell is employed, traces of intermediate compounds with a typical IR absorption band (say, enols) can be detected in the irradiated vapor.

E) *EPR spectroscopy*

This technique is suitable for the detection of free radicals and excited triplet states.

6. Study of reaction mechanisms by isotopic labeling

This method, which is often employed in classical organic chemistry, can be applied to photochemical work as well.

7. Trapping of free radicals formed in the course of primary processes

If a substance which is highly reactive toward free radicals is introduced, the free radicals may become bound before undergoing a secondary reaction. Such substances include paraffins, nitric oxide (NO), olefins, iodine and oxygen. It will be noted that iodine and oxygen often have different effects on the mechanism of a photochemical reaction. The free radical trapping technique must therefore be employed with great caution.

8. Kinetic methods

Kinetic studies of photochemical reactions are very often employed in the determination of their mechanism. Nevertheless, caution is indicated in interpreting the results, since two different mechanisms may often result in an identical kinetic behavior. A simple example of a kinetic study was given in Chapter 3 (Stern-Volmer equation). The reaction rate may be affected by a number of different factors.

A) *Effect of temperature*

If the quantum yield of a reaction product is independent of the temperature, it may usually be concluded that the product is primary. However, this is not invariably so: a primary process may be temperature-dependent and, on the contrary, a process which is independent of the temperature is not necessarily primary.

B) *Effect of light intensity*

Primary quantum yields, unlike secondary quantum yields, are independent of the intensity of the radiation absorbed.

C) *Effect of absorbed wavelength*

This effect may only be apparent, and may in fact be due to the different intensities of the different wavelengths emitted by the light source. If the quantum yield is in fact different at different wavelengths, the reaction takes place prior to the dissipation of excess vibrational energy in the medium. Such an effect can be observed in the vapor phase under reduced pressure; in liquid phase the effect will be exceedingly rare.

D) *Solvent effect*

Such studies mainly concern the behavior of the solvent as proton donor and its viscosity.

9. Sensitization and quenching

Techniques based on electronic energy transfer (Chapter 3) are very useful in determining the nature of photochemical mechanism In particular, they make it possible to determine the nature and the energy of the excited state. Photosensitization trials are particularly suitable for this purpose. Results of quenching experiments are often less reliable.

III. PHOTOCHEMISTRY OF HYDROCARBONS

1. Saturated hydrocarbons /15/

Saturated hydrocarbons absorb in the far UV; their primary photodecomposition results in the liberation of a hydrogen molecule and the formation of a carbene.

$$\text{R—CH}_2\text{—R}' \xrightarrow{h\nu} \text{R—}\ddot{\text{C}}\text{—R}' + \text{H}_2\,.$$

2. Unsaturated hydrocarbons /16–24/

Olefins, and compounds with a C=C bond in general, participate in a number of typical photochemical reactions. Since the excitation of an unconjugated double bond takes place in the far ultraviolet (below 2000 Å), the photochemist will mainly attempt to study conjugated double bonds. α,β-Unsaturated carbonyl compounds will usually react as olefins rather than as carbonyl compounds.

A) *cis-trans Isomerization*

This is the most common rearrangement given by open-chain olefins in solution. The irradiation of a *cis-* or a *trans-*olefin yields a mixture of the two isomers. This type of isomerization has already been mentioned (stilbenes, Chapter 3).

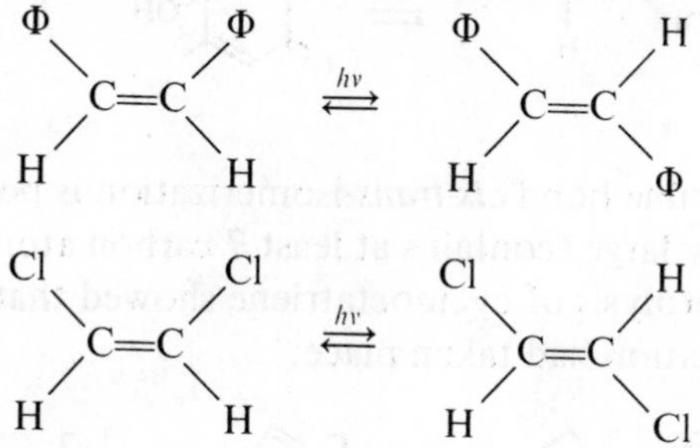

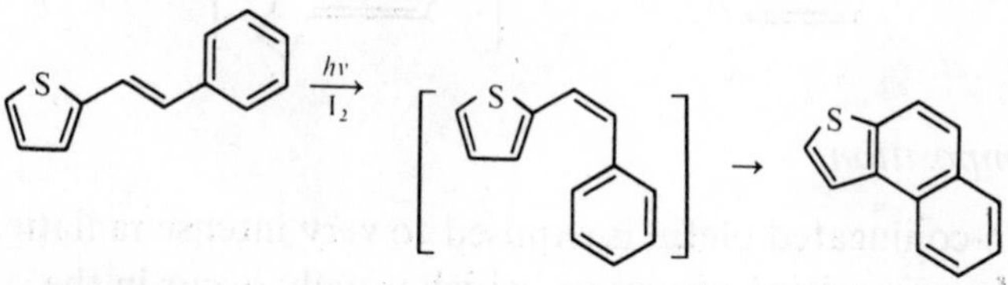

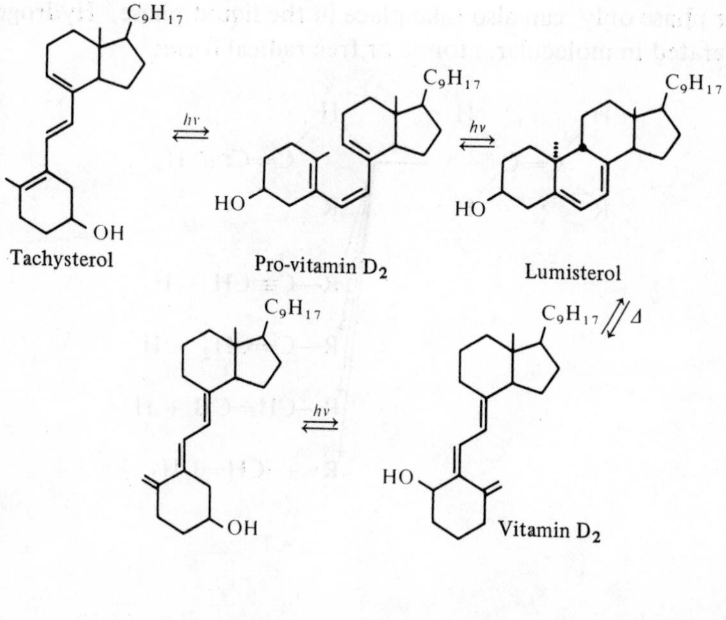

Tachysterol ⇌ (hν) Pro-vitamin D_2 ⇌ (hν) Lumisterol

Vitamin D_2

cis-trans-Isomerization plays an important part in the photochemistry of polyenes, especially in the case of vitamin A and carotenoids. Abscisin II, the regulating hormone of vegetable growth, also undergoes *cis-trans*-isomerization.

$$\xrightleftharpoons{h\nu}$$

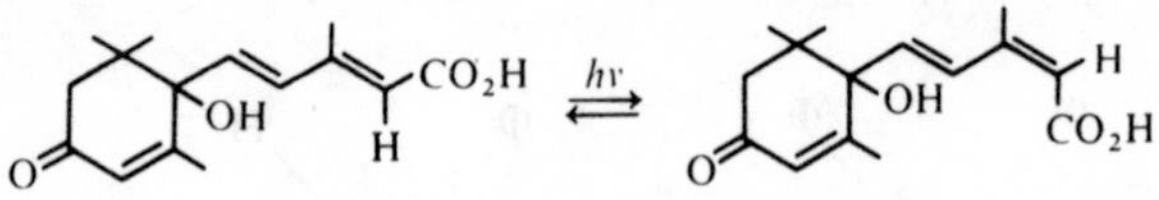

Endocyclic double bond *cis-trans*-isomerization is possible if the ring is sufficiently large (contains at least 7 carbon atoms, cf. Chapter 7). Flash photolysis of cyclooctatriene showed that a double *cis-trans*-isomerization had taken place:

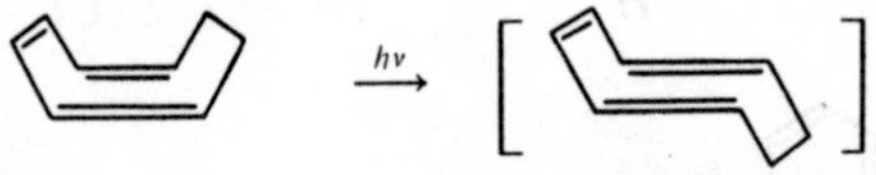

$$\xrightarrow{h\nu} [\quad]$$

B) *Decomposition*

If a non-conjugated olefin is exposed to very intense radiation, primary decomposition processes, which usually occur in the vapor phase only, can also take place in the liquid phase. Hydrogen is liberated in molecular, atomic or free radical form:

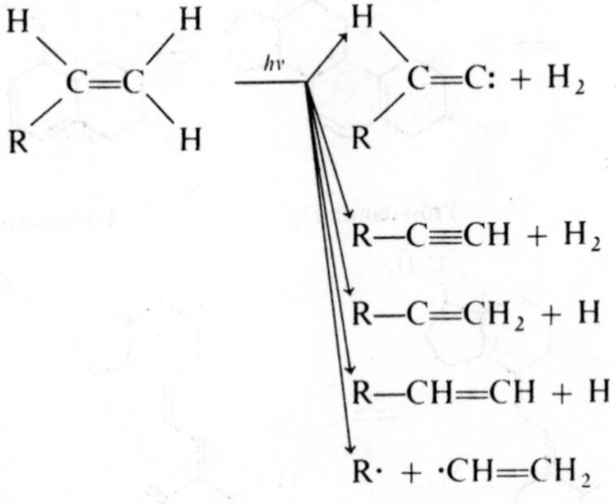

$$(H)(R)C{=}C(H)(H) \xrightarrow{h\nu} (H)(R)C{=}C\!: + H_2$$

$$R{-}C{\equiv}CH + H_2$$

$$R{-}\dot{C}{=}CH_2 + H$$

$$R{-}CH{=}\dot{C}H + H$$

$$R\cdot + \cdot CH{=}CH_2$$

Irradiation of ethylene mainly results in the formation of acetylene and hydrogen.

C) *Rearrangement*

Electrocyclic transformations of polyenes will be discussed in Chapter 6 in more detail. A few examples are given here, along with other types of rearrangement.

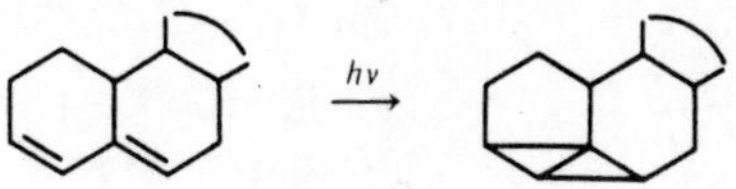

hν

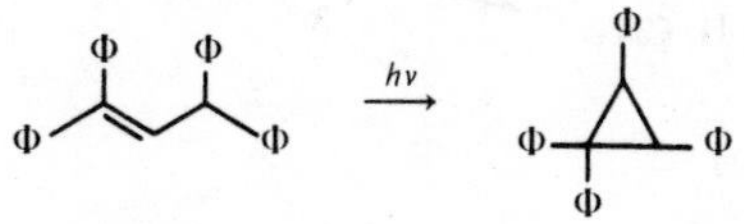

Φ Φ Φ Φ hν Φ Φ Φ Φ

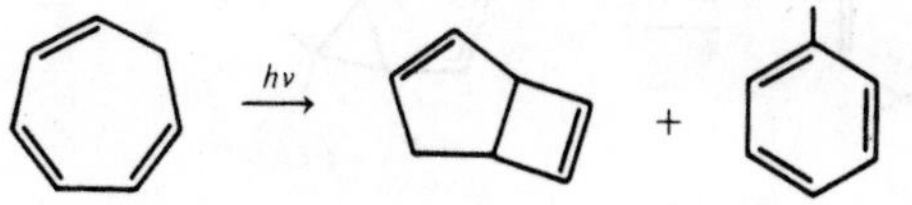

hν +

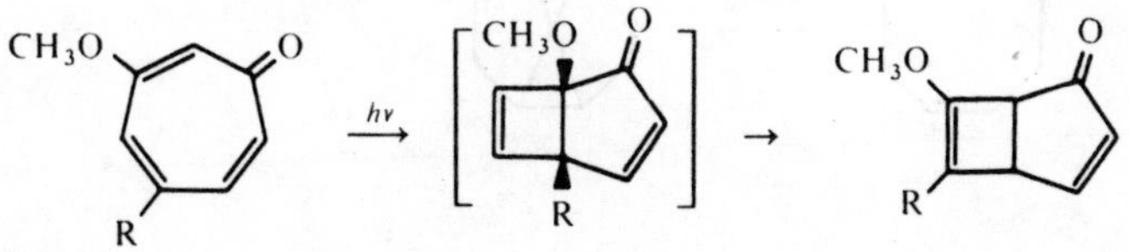

CH_3O O hν CH_3O O R → CH_3O O R

R

($R=CH(CH_3)_2$. Cl)

hν ≡

hν

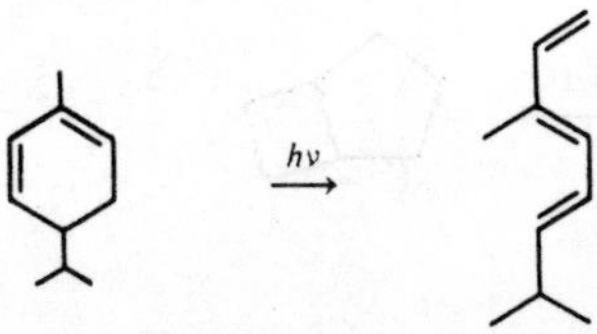

Many sensitized (mostly mercury-sensitized) ring closure reactions are known (R. Srinivasan).

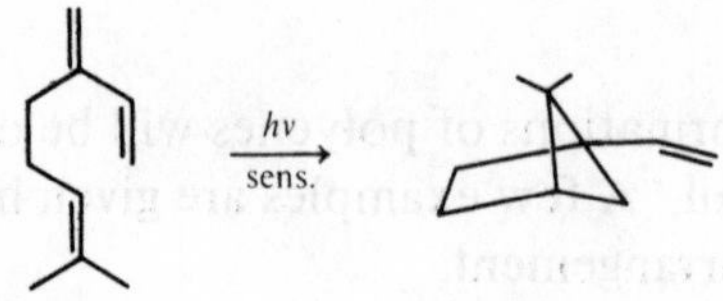

$\xrightarrow[\text{sens.}]{h\nu}$

Myrcene

$\xrightarrow[\text{sens}]{h\nu}$

(R=H, CH_3)

R R R R

$\xrightarrow{h\nu/Hg}$

$\xrightarrow{h\nu/Hg}$

R $\xrightarrow{h\nu/Hg}$ R + R

(R=H, CH_3)

$\xrightarrow{h\nu/Hg}$

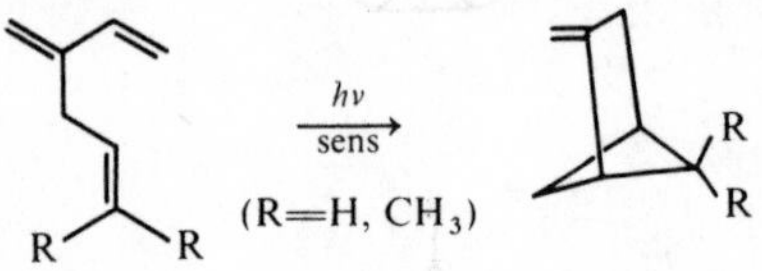

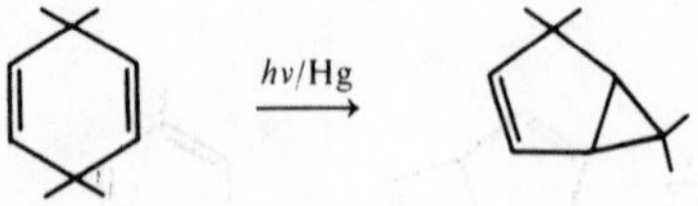

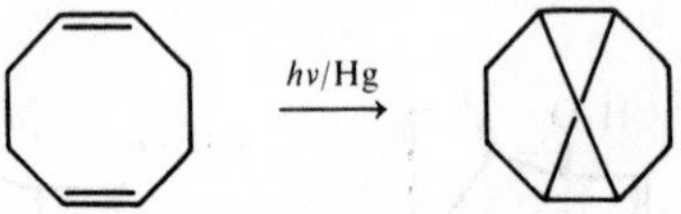

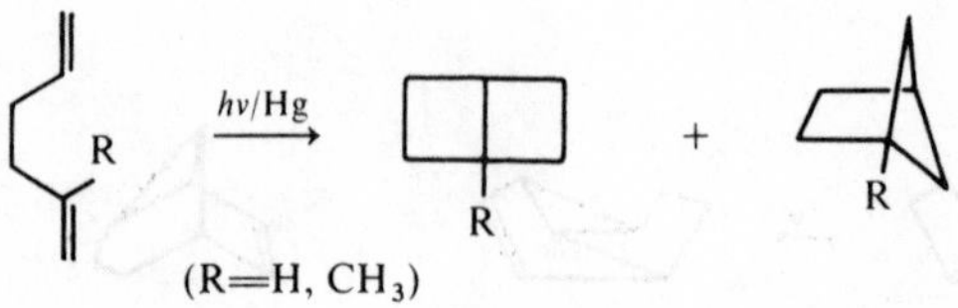

$\xrightarrow{h\nu/Hg}$ (R, R′=H, CH_3)

(R=H, R′=H, CH_3)

$\xrightarrow{h\nu/Hg}$ ($R=CH_3$; R,R=⬡) $h\nu/Hg$

D) *Intermolecular addition*

a) *Dimerization* /25, 26/. The formation of a cyclobutane derivative by dimerization of a double bond will also be discussed in Chapter 6.

$\xrightarrow{h\nu}$

Many α, β-unsaturated carbonyl compounds give cyclobutane dimers.

b) *Various addition reactions.* The addition of α, β-unsaturated carbonyl compounds to olefins will be dealt with in Chapter 7.

Cyclobutane derivatives are formed as in dimerizations:

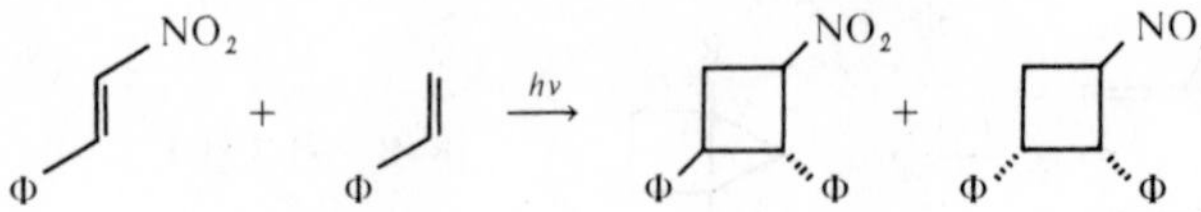

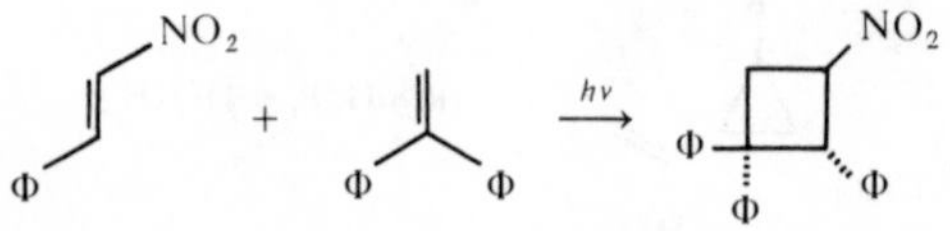

3. Aromatic hydrocarbons

Benzene will be discussed in Chapter 5.

Dimerization is an important photochemical reaction given by polycyclic aromatic hydrocarbons:

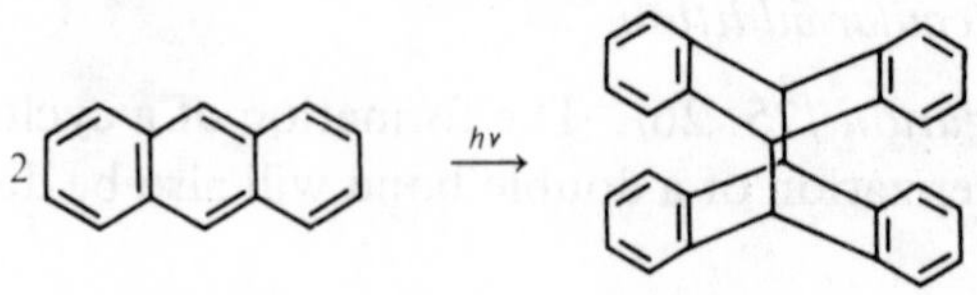

Alkylbenzenes undergo decomposition to yield benzyl radicals:

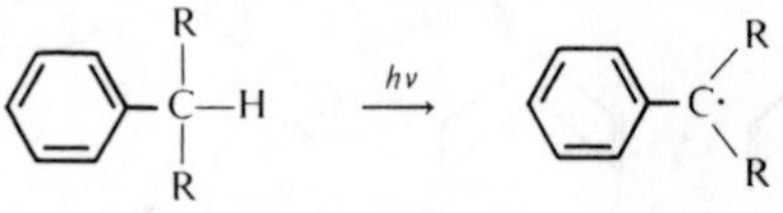

The substitution of aromatic derivatives, which proceeds only with great difficulty in the ground state, takes place much more readily in the excited state. Thus, nitrophenyl phosphates and sulfates, which are very stable in aqueous solutions in the dark, are rapidly hydrolyzed when illuminated: the *m*-derivatives are 20–30 times more reactive than the *o*- and *p*-derivatives, contrary to what is observed in the ground state. It has been established by Havinga /27–29/ that the charge distribution in the excited state is often responsible for the selectivity of the reaction.

$$m\text{-}NO_2C_6H_4OPO_3^{2-} \xrightarrow[ROH]{h\nu} m\text{-}NO_2C_6H_4OH + RPO_4^{--}$$

$$m\text{-}NO_2C_6H_4OPO_3^{2-} \xrightarrow[CH_3NH_2]{h\nu} m\text{-}NO_2C_6H_4NHCH_3 + PO_4H^{--}$$

$$m\text{-}NO_2C_6H_4OCH_3 \xrightarrow[CH_3NH_2]{h\nu} m\text{-}NO_2C_6H_4NHCH_3 + CH_3OH$$

Other substitution reactions, which have not been observed to take place in the dark, and which are directed according to thermal substitution rules, have been reported to proceed by way of intermediate complexes:

$$C_6H_5NO_2 \xrightarrow[NH_3\ liq.]{h\nu} NH_2\text{-}C_6H_4\text{-}NO_2$$

4. Halogenated hydrocarbons /30, 31/

The photodissociation of an alkyl halide R–X into an alkyl radical and a free halogen atom is very efficient ($\varphi \sim 1$) if X=Cl, Br, I. The absorption wavelength varies with the electronegative character and the number of halogen atoms in the molecule; thus, λ_{max} is about 1730 Å for methyl chloride, 2576 Å for methyl iodide, and 3490 Å for iodoform.

$$R\text{—}X \xrightarrow{h\nu} R\cdot + X\cdot \,.$$

Another primary process may also take place:

$$R_2CHCH_2X \xrightarrow{h\nu} R_2C{=}CH_2 + HX \,.$$

If a hydrocarbon molecule is substituted by two different halogen atoms, the weaker C–X bond is preferentially broken:

$$Cl_2CH\text{—}Br \xrightarrow{h\nu} Cl_2CH\cdot + .Br\ .$$

Irradiation of aromatic halogenated hydrocarbons (X=Cl, Br, I) results in the elimination of the halogen atom:

$$C_6H_5Br \xrightarrow{h\nu} C_6H_5\cdot + Br\cdot$$

This elimination is particularly efficient from iodo derivatives; the intermediate aromatic free radical can participate in very interesting syntheses (N. Kharash):

$$p\text{-}I\text{-}C_6H_4\text{-}CHO \xrightarrow{h\nu} \cdot C_6H_4\text{-}CHO \xrightarrow{C_6H_6} p\text{-}C_6H_5\text{-}C_6H_4\text{-}CHO \quad 90\ \%$$

$$C_6H_5I \xrightarrow{h\nu} C_6H_5\cdot \xrightarrow{P(OR)_3} C_6H_5P(O)(OR)_2 + RI$$

$$CH_3O_2C\text{-}C_6H_4\text{-}C(NO_2)\text{=}CH\text{-}C_6H_4\text{-}I \xrightarrow{h\nu} \text{(phenanthrene, } CH_3O_2C\text{, } NO_2\text{)} \quad 50\ \%$$

IV. PHOTOCHEMISTRY OF CARBONYL COMPOUNDS

Aldehydes and ketones will be discussed in Chapter 7. The reactions of carboxylic acids, esters, anhydrides, acid chlorides, and acid amides will be dealt with here.

1. Carboxylic acids

Monobasic, and probably also polybasic carboxylic acids decompose with evolution of CO or CO_2:

$$R\text{—}CO_2H \xrightarrow{h\nu} R\cdot + CO_2H\cdot \longrightarrow CO_2 + H\cdot$$

$$\longrightarrow H\cdot + RCO_2\cdot \longrightarrow CO_2 + R\cdot$$

$$CO_2 + H\cdot \text{ and } CO_2 + R\cdot \longrightarrow R\cdot + H\cdot + CO_2$$

$$\longrightarrow OH\cdot + RCO\cdot \longrightarrow R\cdot + CO$$

$$\longrightarrow RH + CO_2$$

$$CH_2(CO_2H)_2 \xrightarrow{h\nu(\lambda < 2\,500\text{ Å})} CH_3\text{—}CO_2H + CO_2\,.$$

Acids with double bonds undergo *cis-trans*-isomerization (see photochemistry of olefins) in preference to decomposition. α-Keto acids decompose with conversion to aldehydes:

$$CH_3\text{—}\underset{\underset{O}{\|}}{C}\text{—}CO_2H \xrightarrow{h\nu} [\text{cyclic H-bonded intermediate}] \rightarrow CH_3\text{—}CHO + CO_2$$

HO_2C, OH, OH, CO_2H

O, CO_2H $\xrightarrow[CH_3CHOHCH_3]{h\nu}$

$\xrightarrow[H_2O]{h\nu}$ CHO $+ CO_2$

2. Carboxylic acid esters

Esters of saturated monobasic fatty acids give the same types of reactions as those observed for the free acids. If a hydrogen atom is available in the γ-position to the carbonyl group, the molecule is cleaved (Chapter 7); the hydrogen atom can be situated in the γ-position to the carboxyl group of the acid, or in the β-position to the hydroxyl group of the ester:

R, R, H, O, OR′ $\xrightarrow{h\nu}$ $R_2CH{=}CH_2 + CH_3{-}C({=}O){-}OR'$

R′, H, R′, O, R, O $\xrightarrow{h\nu}$ $R{-}C({=}O){-}OH + R'_2CH{=}CH_2$

α, β-Unsaturated esters in solution undergo *cis-trans*-isomerization, and also cyclo addition reactions. The synthesis of α-bourbonene is a photochemical intramolecular cycloaddition:

CO_2Et, H, CO_2Et $\xrightarrow{h\nu}$ CO_2Et, H, H, H, CO_2Et $\rightarrow$ O, H, H, H, O

tBuO$^\ominus$

tBuO$^{\ominus}$

$LiAlH_4/AlCl_3$

(±) α-bourbonene

Aromatic esters undergo a photochemical Fries type rearrangement, which is different from the acid-catalyzed Fries rearrangement /32/:

$h\nu$ 140° $AlCl_3$

Lactones react at the same time both as esters and as the corresponding cyclic ketones:

+ $h\nu$ ⟶

$CH_3—CH_2—CHO$ + CO

cyclopropane + CO_2

$CH_3—CHO$ + $CH_2=C=O$

3. Carboxylic acid anhydrides and acid chlorides

Dissociation of the molecule into free radicals is a major reaction

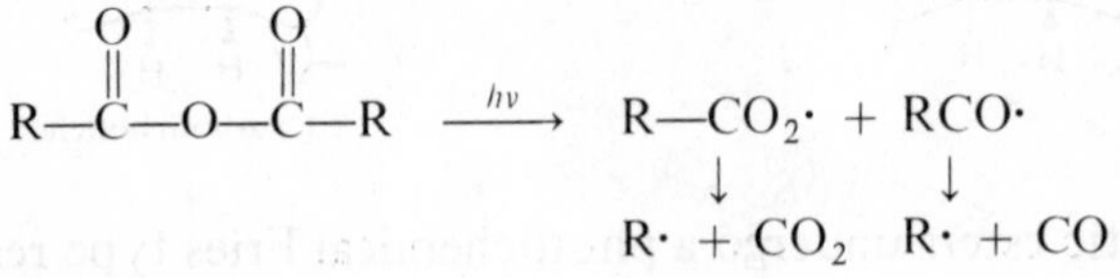

$$R-\overset{\overset{\displaystyle O}{\|}}{C}-O-\overset{\overset{\displaystyle O}{\|}}{C}-R \xrightarrow{h\nu} R-CO_2\cdot + RCO\cdot$$

$$R-CO_2\cdot \rightarrow R\cdot + CO_2 \qquad RCO\cdot \rightarrow R\cdot + CO$$

but cleavage of a hydrogen atom in the γ-position to the carbonyl is also observed:

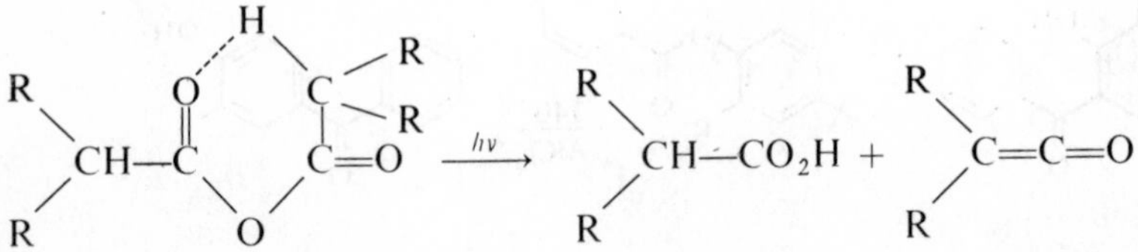

$$R_2CH-\overset{\overset{\displaystyle O}{\|}}{C}-O-\overset{\overset{\displaystyle O}{\|}}{C}-CHR_2 \xrightarrow{h\nu} R_2CH-CO_2H + R_2C=C=O$$

as is cycloaddition of α, β-unsaturated anhydrides:

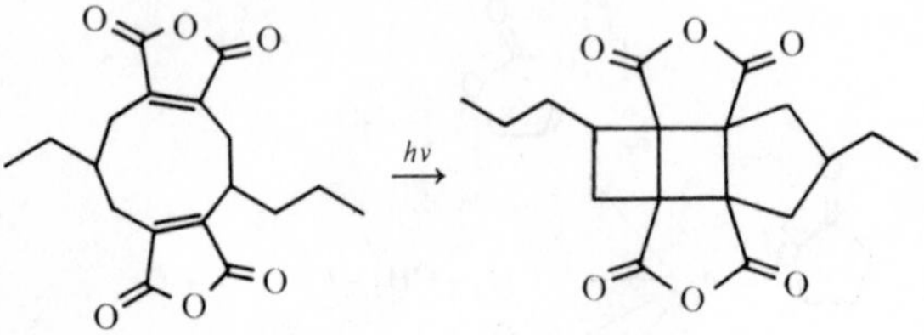

$$\xrightarrow{h\nu}$$

4. Carboxylic acid amides

Molecules of carboxylic acid amides decompose in a manner which is similar to that of other carbonyl compounds: by dissociation into free radicals and by a type II Norrish reaction, as well as by characteristic dissociations which do not involve free radicals:

$$R{-}\underset{\underset{O}{\|}}{C}{-}NH_2 \xrightarrow{h\nu} NH_2\cdot + RCO\cdot \longrightarrow R\cdot + CO$$
$$\longrightarrow R\cdot + \cdot CONH_2 \longrightarrow CO + NH_2\cdot$$
$$\longrightarrow RCN + H_2O$$
$$\longrightarrow RNH_2 + CO$$

$$NH_2{-}\overset{\overset{O\cdots H}{\|}}{C}{-}CH_2{-}CH_2{-}CR_2H \xrightarrow{h\nu} NH_2{-}\overset{\overset{O}{\|}}{C}{-}CH_3 + R_2C{=}CH_2\,.$$

The $\cdot CONH_2$ radical formed in a primary process may add on to olefins, as was shown by Elad /34/ in the case of formamide:

$$H{-}\underset{\underset{O}{\|}}{C}{-}NH_2 \xrightarrow{h\nu} H\cdot \quad \cdot\underset{\underset{O}{\|}}{C}{-}NH_2 \xrightarrow{R-CH=CH_2} R{-}CH_2{-}CH_2{-}CONH_2$$

$$R{-}CH{=}CH{-}CO_2Me \xrightarrow[h\nu]{HCONH_2} R{-}\underset{\underset{CONH_2}{|}}{CH}{-}CH_2{-}CO_2Me$$

Barton /33/ studied the photochemistry of *N*-iodoamides. These compounds, which are prepared by the action of lead tetraacetate and iodine, or of *tert*-butyl hypochlorite and iodine, are photolyzed. The resulting free radical withdraws the proton in the γ-position, the final product being a γ-lactone:

$Pb(OAc)_4$, I_2 → NHI; $h\nu$ → NH· + I·

18-Hydroxyoestrone has been prepared in this manner /33/:

$Pb(OAc)_4 + I_2$, $h\nu$

$Pb(OAc)_4 + I_2$, $h\nu$

1. Li
2. $LiAlH_4$
3. Ac_2O

$SOCl_2$—Py

1. O_3
2. OH^-

V. PHOTOCHEMISTRY OF ALCOHOLS AND ETHERS /35/

Alcohols and saturated ethers absorb in the UV, below 2000 Å. When irradiated at these wavelengths, they undergo a free radical or a molecular decomposition:

$$R{-}CH_2{-}CH_2OH \xrightarrow{h\nu} \begin{cases} OH\cdot + RCH_2CH_2\cdot \rightarrow R\cdot + CH_2{=}CH_2 \\ H\cdot + RCH_2CH_2O\cdot \rightarrow RCH_2\cdot + CH_2{=}O \\ RCH{=}CH_2 + H_2O \\ RCH_2CHO + H_2 \\ RCH_3 + CH_2{=}O \end{cases}$$

$$R{-}CH_2{-}CH_2{-}O{-}CH_2{-}CH_2{-}R \xrightarrow{h\nu} \begin{cases} R{-}CH_2{-}CH_2{-}O\cdot + \cdot CH_2{-}CH_2{-}R \\ \quad \downarrow \qquad\qquad\qquad \downarrow \\ RCH_2\cdot + CH_2{=}O \quad CH_2{=}CH_2 + R\cdot \\ RCH{=}CH_2 + HO{-}CH_2{-}CH_2{-}R \\ R{-}CH_2{-}CH{=}O + CH_3{-}CH_2{-}R \end{cases}$$

Peroxides absorb at wavelengths longer than 2000 Å with rupture of the O–O bond; photolysis of peroxides is a good method of preparation of RO· radicals:

$$ROOR' \xrightarrow{h\nu} RO\cdot + R'O\cdot\,.$$

A photochemical Claisen rearrangement was observed for a number of aromatic ethers in hydrogen-donating solvents. It is believed that the *p*-isomer alone is formed:

$$C_6H_5{-}O{-}C_6H_5 \xrightarrow[ROH]{h\nu} C_6H_5{-}OH + C_6H_5{-}C_6H_4{-}OH$$

$$C_6H_5{-}O{-}CH_2{-}C_6H_5 \xrightarrow[ROH]{h\nu} C_6H_5{-}OH + C_6H_5{-}CH_2{-}C_6H_4{-}OH$$

Dihydrofurane ethers are converted to ketones by ring opening:

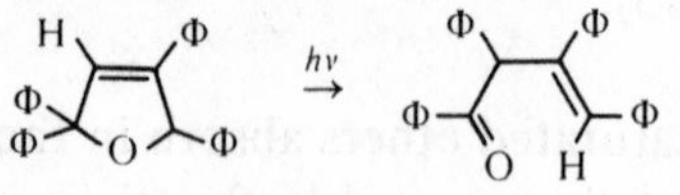

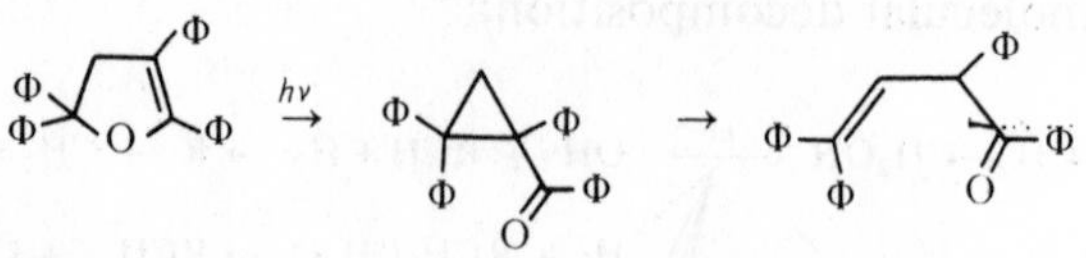

VI. PHOTOCHEMISTRY OF NITROGEN COMPOUNDS /36–43/

1. Amines

The main schemes of photolytic dissociation of the various types of amines are as follows:

primary amines:

$$R{-}NH_2 \xrightarrow{h\nu} RNH\cdot + H\cdot ;$$

secondary amines:

$$R_2NH \xrightarrow{h\nu} R_2N\cdot + H\cdot$$

$$R_2NH \xrightarrow{h\nu} R\cdot + R\dot{N}H$$

This process was observed in the photolysis of N-methylactinomycin C_2:

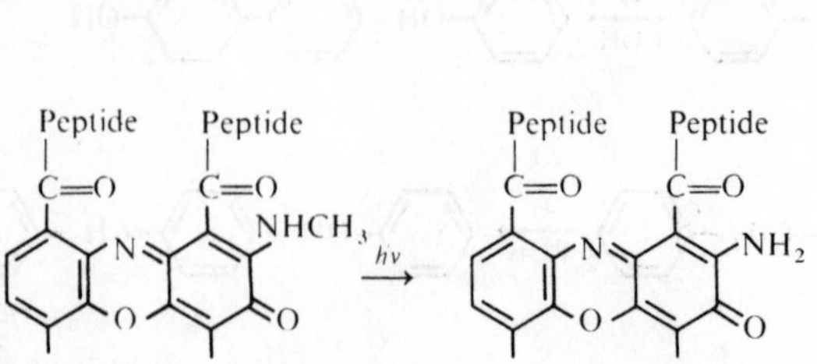

tertiary amines:

$$R_3N \xrightarrow{h\nu} R\cdot + R_2N\cdot$$

aromatic amines: these compounds are much more stable than aliphatic amines and are decomposed with difficulty.

Chloramines give the Hofmann-Löffler-Freytag reaction:

$$\text{Cl CH}_3\text{ N} \xrightarrow{h\nu} \text{CH}_3\text{ N}$$

The partial synthesis of an alkaloid – demissidine – has been realized in this manner:

Demissidine

2. Azo derivatives

In the vapor phase, aliphatic azo derivatives decompose into free radicals:

$$R{-}N{=}N{-}R \xrightarrow{h\nu} R\cdot + \cdot N_2R$$

$$\cdot N_2R \longrightarrow R\cdot + N_2$$

In the liquid phase, both aliphatic and aromatic azo derivatives undergo *syn-anti*-isomerization:

$$\Phi-\ddot{N}=\ddot{N}-\Phi \ (anti) \underset{}{\overset{h\nu}{\rightleftarrows}} \Phi-\ddot{N}=\ddot{N}-\Phi \ (syn)$$

This type of isomerization has also been observed for imines:

$$C_6H_5-CH=\ddot{N}-C_6H_5 \overset{h\nu}{\rightleftarrows} C_6H_5-CH=\ddot{N}-C_6H_5$$

3. Diazo compounds

Diazo compounds are readily photolyzed into molecular nitrogen and carbenes, the latter being conveniently prepared by this method:

$$R_2C=\overset{\oplus}{N}=\overset{\ominus}{N} \xrightarrow{h\nu} R_2C: + N_2 .$$

The carbenes thus formed participate in secondary reactions (G. Cauquis):

$$\text{10-diazoanthrone } (C=N^{\oplus}=N^{\ominus}) \xrightarrow{h\nu} \text{anthrone carbene } (\ddot{C}) \xrightarrow{CH_2=CH_2} \text{spiro-cyclopropane anthrone}$$

$$\downarrow RH$$

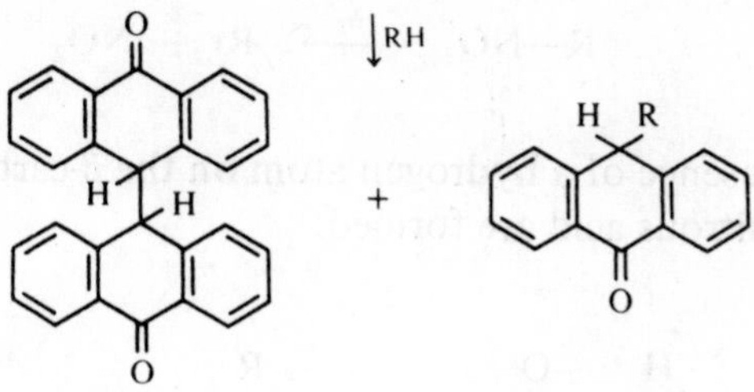

RH

H R

H H

+

O

O

O

Diazoketones are also converted to carbenes which usually rearrange to yield ketenes /36/:

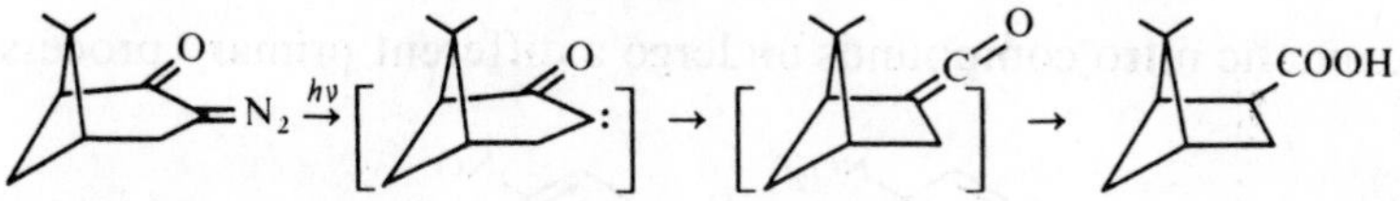

O

$=N_2 \xrightarrow{h\nu}$ [O :] → [C=O] → COOH

The photolysis of diazoketones has also been utilized in the synthesis of nor-A-steroids and nor-D-steroids:

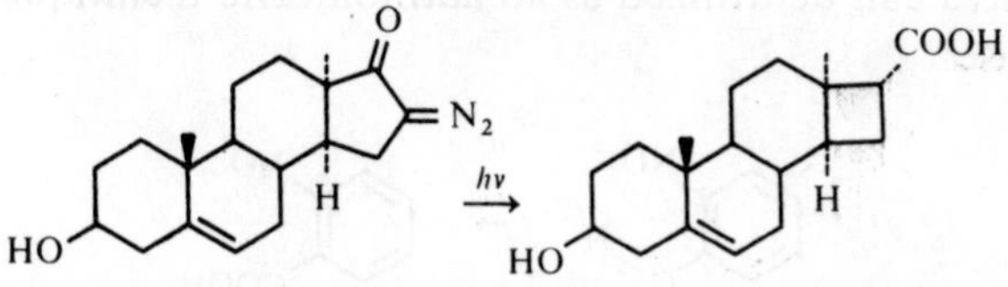

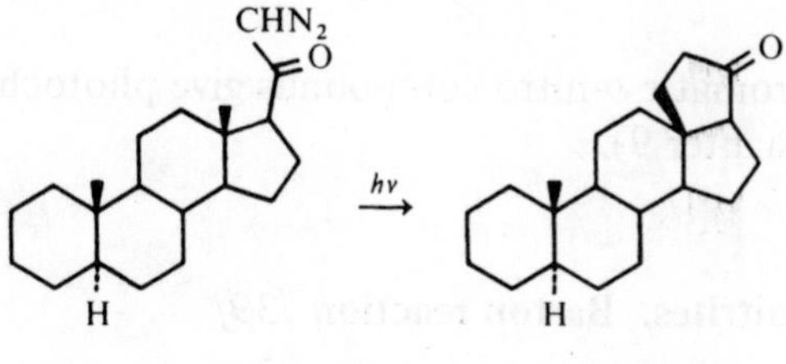

O

$=N_2$

HO H $\xrightarrow{h\nu}$ HO H

COOH

CHN_2

O

$\xrightarrow{h\nu}$

H H

O

4. Nitro compounds /37/

Aliphatic nitro compounds are primarily decomposed by light into free radicals:

$$R{-}NO_2 \xrightarrow{h\nu} R\cdot + \cdot NO_2\,.$$

In the presence of a hydrogen atom on the β-carbon atom, an olefin and nitrous acid are formed:

$$R_2C(H){-}CH_2{-}N(=O){\rightarrow}O \xrightarrow{h\nu} R_2C{=}CH_2 + NO_2H\,.$$

Aromatic nitro compounds undergo a different primary process:

$$C_6H_5NO_2 \xrightarrow{h\nu} C_6H_5NO$$

The rearrangement of *o*-nitrobenzaldehyde to the corresponding nitroso acid can be utilized as an actinometric technique in the solid phase:

$$o\text{-}NO_2C_6H_4CHO \xrightarrow{h\nu} o\text{-}NOC_6H_4COOH$$

Certain aromatic *o*-nitro compounds give photochromic reactions (see Chapter 9).

5. Organic nitrites. Barton reaction /39/

The primary process involved in the photolysis of nitrites is the homolytic cleavage into the alcoholate* radical RO· and ·NO; the quantum yield of this process is usually close to unity:

* It will be noted that the photolysis of organic hypochlorites R–OCl yields the same RO· radical; thus, the photolytic reaction is the same as that given by nitrites.

$$R{-}O{-}NO \xrightarrow{h\nu} RO\cdot + \cdot NO\,.$$

The alcoholate radical thus formed can participate in several different reactions:

recombination: $RO\cdot + \cdot NO \rightarrow RONO$

dimerization: $RO\cdot + \cdot OR \rightarrow ROOR$

addition to an unsaturated compound: $>C{=}C< \xrightarrow{RO\cdot} >\dot{C}{-}C(OR)<$

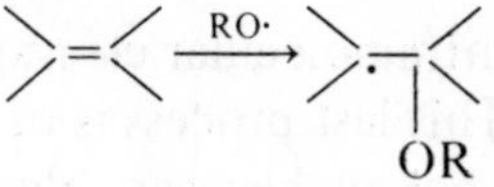

dismutation:

$$2\,RCH_2O\cdot \rightarrow RCH_2OH + RC(=O)H$$

addition of hydrogen from the solvent or from another molecule:

$$RO\cdot \xrightarrow{R'H} ROH + R'\cdot$$

decomposition into smaller radicals and molecules:

$$RCH_2O\cdot \rightarrow RCHO + H\cdot$$

$$RCH_2O\cdot \rightarrow R\cdot + CH_2O$$

The possible rearrangements include:

ring opening:

O· H → CHO $\dot{C}H_2$

ring closure:

OH $\dot{C}H_2$ O → HO O

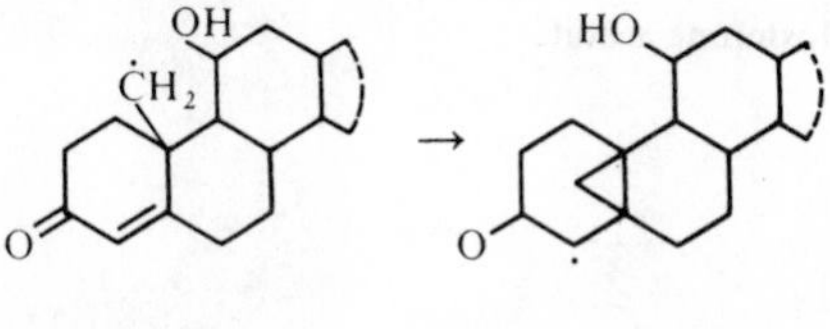

ring expansion:

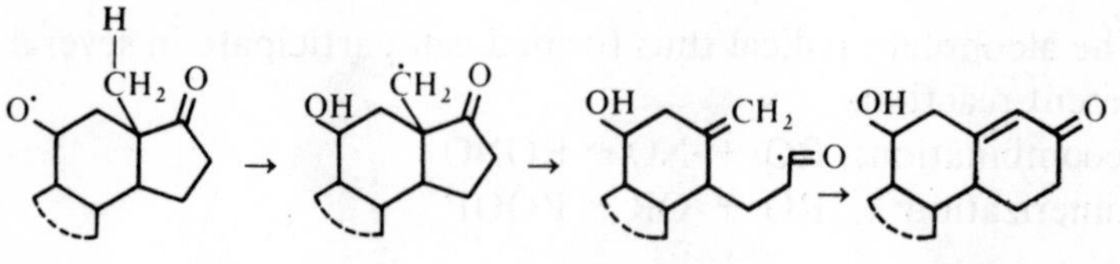

intramolecular cleavage of a hydrogen atom (Barton's reaction).

This last process is of the greatest interest to the organic chemist, since it enables very elegant syntheses to be carried out. An example is the synthesis of aldosterone which was realized by Barton.

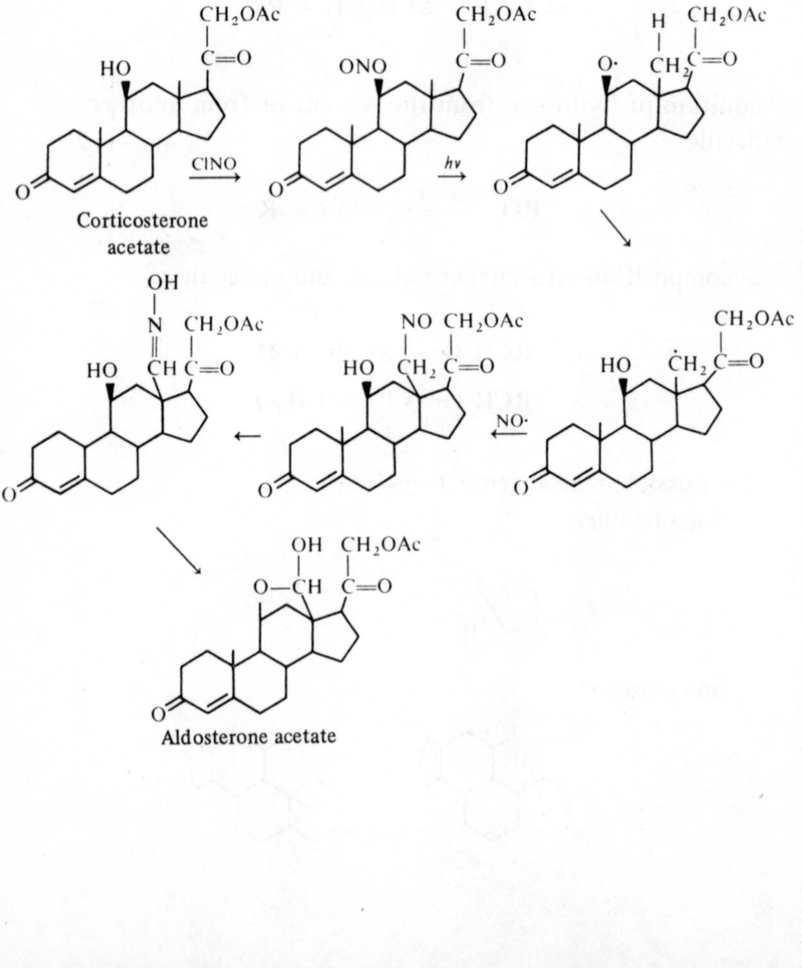

ClNO

$h\nu$

Corticosterone acetate

NO·

Aldosterone acetate

19-Nor-corticoids can be obtained in a similar manner:

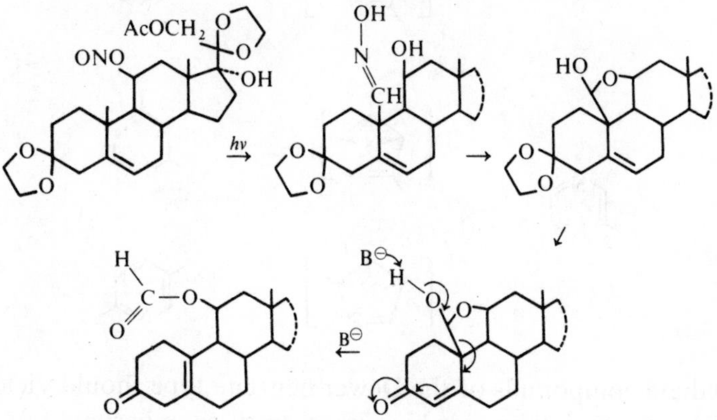

6. Heterocyclic nitrogen compounds /40–41/

According to recent results, aromatic heterocyclic nitrogen compounds such as pyridine or pyrazine undergo reversible isomerizations, similarly to benzene (Chapter 5). This isomerization is known with certainty to take place in the case of pyrazine, since the rearrangement of the instable isomer yields pyrimidine, in addition to the original pyrazine. Studies of substituted pyrazines indicate that the intermediate isomer is of the benzvalene type:

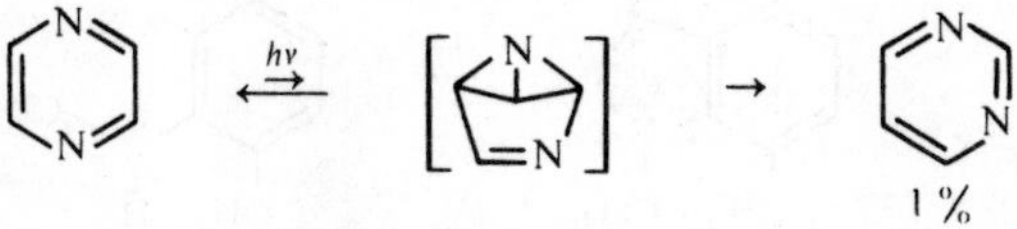

In fact, 2, 6-dimethylpyrazine and 2, 5-dimethylpyrazine yield only dimethylpyrimidines originating from an intermediate product of benzvalene type. As in the case of benzene, this intermediate compound is assumed to originate from the singlet state ($\pi \rightarrow \pi^*$) of pyrazines.

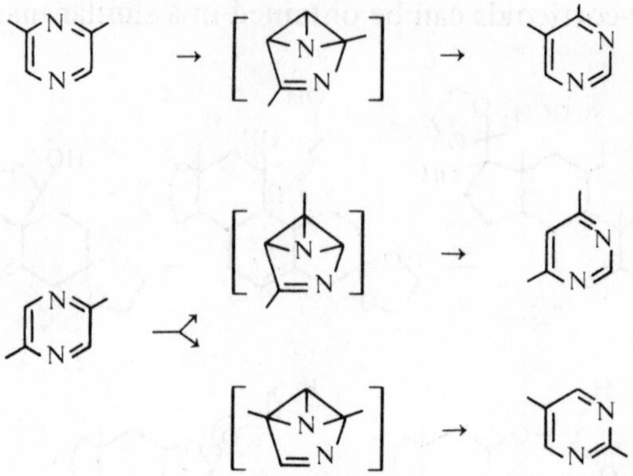

Intermediate compounds of the Dewar benzene type should yield other dimethylpyrimidines, which is contrary to experiment (N. Ivanoff).

2-Alkyl-1H-indazoles are quantitatively photoisomerized to benzimidazoles:

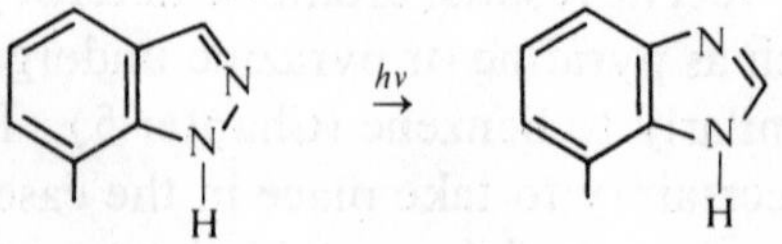

The corresponding N-methyl compound behaves in a different manner:

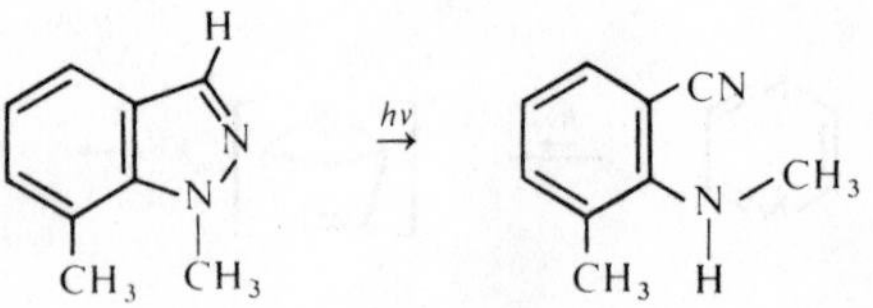

The course of the photochemical rearrangement of pyrimidines could be clarified:

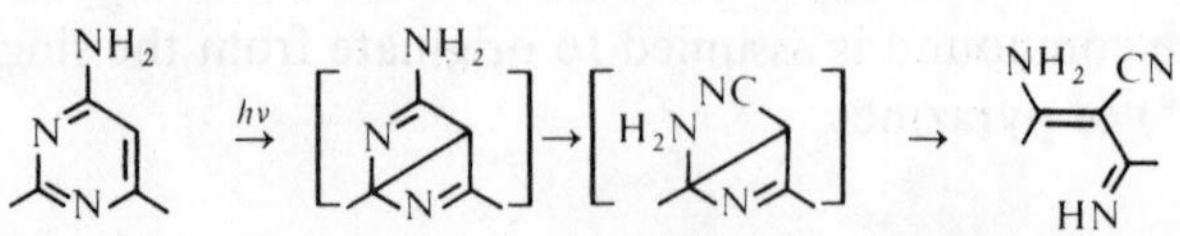

An extensive account of the photochemistry of heterocyclic nitrogen compounds is outside the scope of this chapter. A few examples are given below.

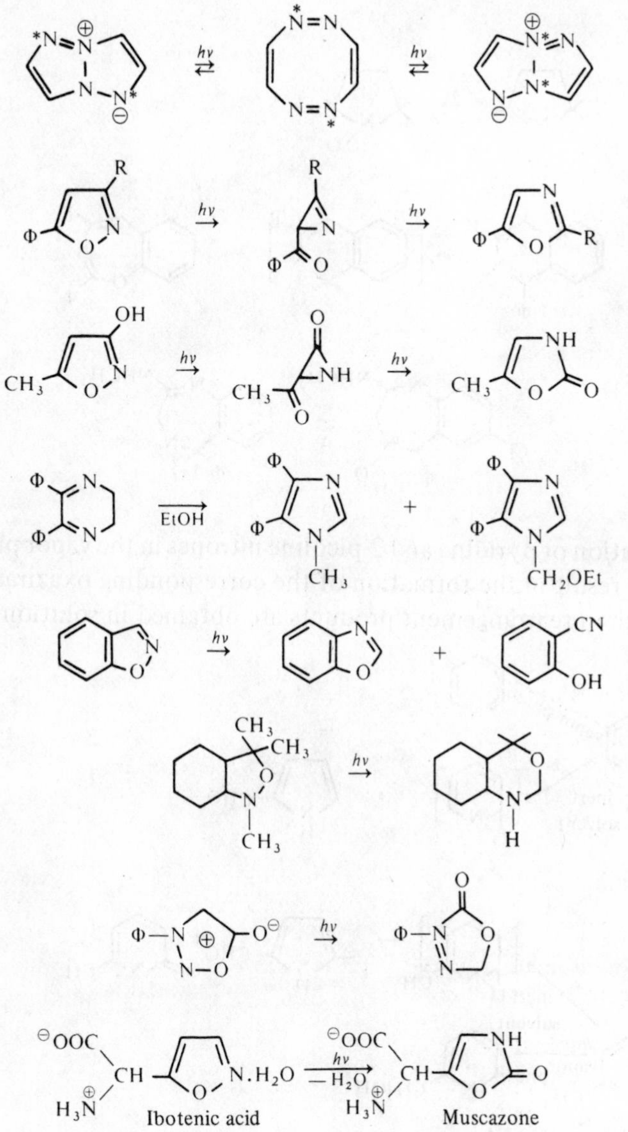

Ibotenic acid → Muscazone

7. Nitrones

Photochemical cyclization of nitrones yields oxazirans, which may in turn undergo a rearrangement or else revert to the initial nitrone.

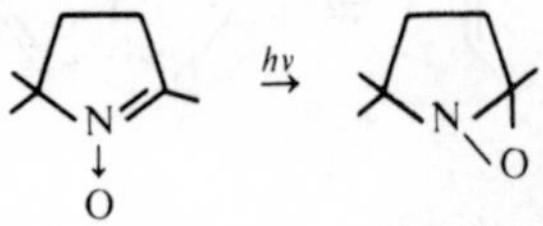

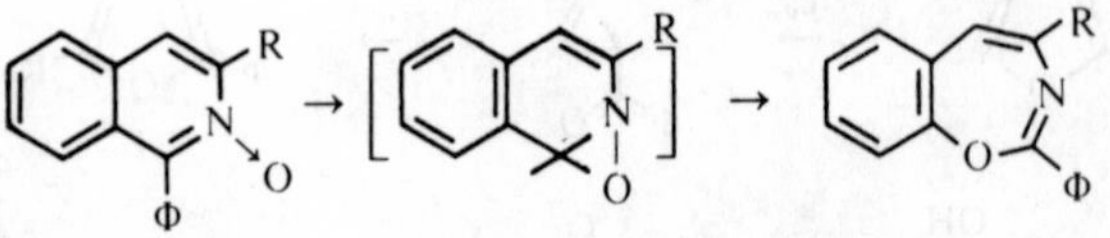

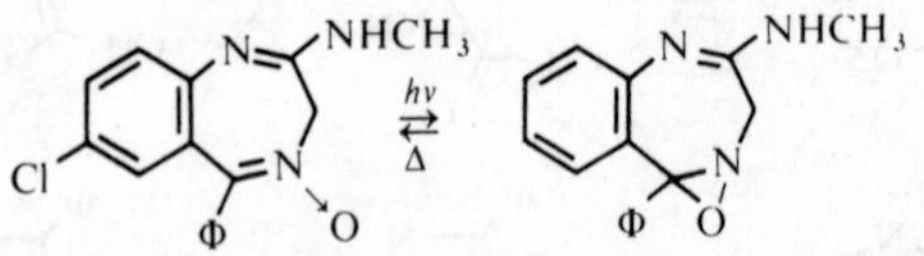

Irradiation of pyridine and 2-picoline nitrones in the vapor phase does not result in the formation of the corresponding oxazirans, but oxaziran rearrangement products are obtained in solution.

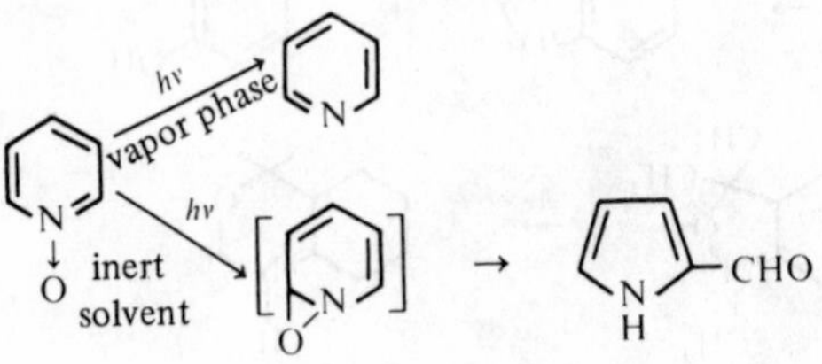

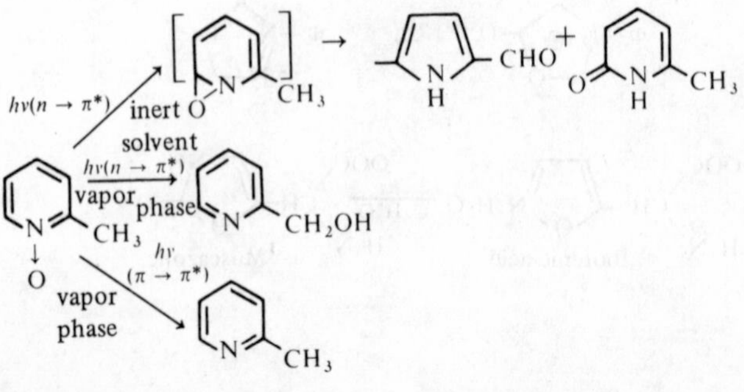

8. Azides

Photochemical decomposition of azides is accompanied by the evolution of nitrogen; the free radical formed generally undergoes intramolecular cyclization:

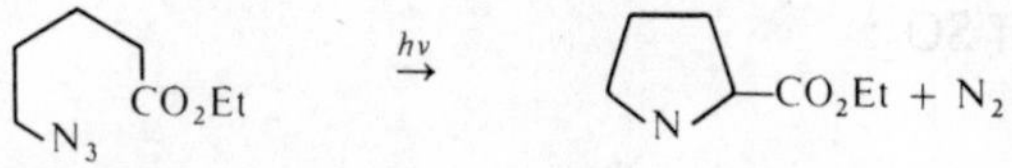

CO_2Et N_3 $\xrightarrow{h\nu}$ N —CO_2Et + N_2

Acid azides react in the same manner, yielding lactams:

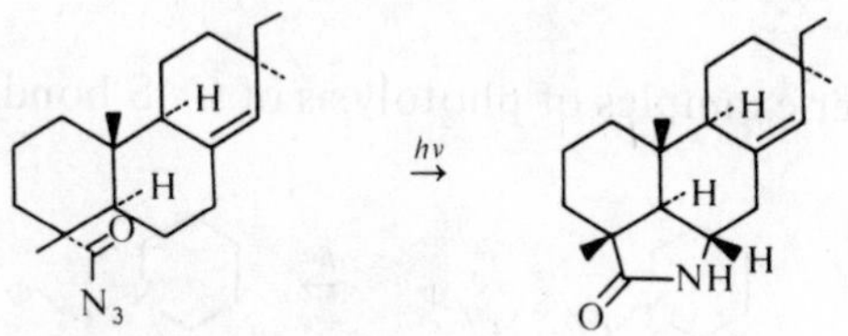

H H O N_3 $\xrightarrow{h\nu}$ H H H NH H O

VII. PHOTOCHEMISTRY OF SULFUR COMPOUNDS /38/

The main primary process in the photolysis of mercaptans is

$$\mathrm{RSH} \xrightarrow{h\nu} \mathrm{RS\cdot} + \mathrm{H\cdot}\ .$$

The RS· radical can add onto a double bond:

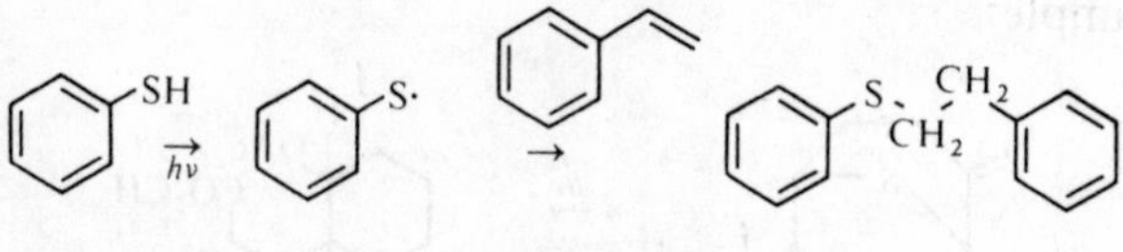

SH $\xrightarrow{h\nu}$ S· $\rightarrow$ S CH_2 CH_2

The photolysis of sulfides takes a similar course:

$$\mathrm{R{-}S{-}R} \xrightarrow{h\nu} \mathrm{RS\cdot} + \mathrm{\cdot R}\ .$$

Disulfides are selectively photolyzed with rupture of the S–S bond:

$$R—S—S—R \xrightarrow{h\nu} 2\,RS\cdot$$

Aromatic sulfones undergo photochemical decomposition with liberation of SO_2:

$$\text{(1,3-diphenyl-1,3-dihydrobenzo[c]thiophene 2,2-dioxide)} \xrightarrow{h\nu} \text{(1,2-diphenylbenzocyclobutene)}$$

Many other examples of photolysis of C–S bonds are known:

$$\text{C}_5\text{H}_{10}\text{N—C(=S)—S—C(=O)—}\Phi \xrightarrow{h\nu} \text{C}_5\text{H}_{10}\text{N—C(=O)—}\Phi$$

$$R—C(=O)—\overset{\ominus}{C}H—\overset{\oplus}{S}(\rightarrow O)(CH_3)—CH_3 \xrightarrow[R'OH]{h\nu} (R)(H)C{=}C{=}O \rightarrow R—CH_2—C(=O)—OR'$$

However, in certain cases it is not the C–S bond but the C–C bond which is photolyzed. Ketones with a β-type sulfur bridge are an example:

$$\text{(bicyclic S-bridged ketone, OAc)} \xrightarrow{h\nu} \text{(thiane with H, OAc, } CO_2CH_3\text{, H)}$$

Thioketones are reduced in a manner similar to ketones, but are not photodecomposed:

$$\text{Ph}_2\text{C=S} \xrightarrow[\text{EtOH}]{h\nu} \text{Ph}_2\text{CH(SH)} + \Phi_2\text{CH—S—S—CH}\Phi_2$$

The photochemical rearrangement of certain heterocyclic sulfur compounds has also been studied:

$$\text{2-}CH_3\text{-5-}\Phi\text{-thiophene} \xrightarrow{h\nu} \text{2-}CH_3\text{-4-}\Phi\text{-thiophene}$$

The mechanism of this rearrangement may involve one of two alternative intermediate compounds, but the identity of the compound has not yet been conclusively established.

The same kind of rearrangement has been observed in the case of 2, 2′-dithiophene:

$$\text{2,2'-dithiophene} \xrightarrow{h\nu} \text{2,3'-dithiophene} \xrightarrow{h\nu} \text{3,3'-dithiophene}$$

Bibliography

1. Kan, R. O. Organic Photochemistry. McGraw-Hill Company, N. Y. 1966.
2. Schönberg, A. Preparative Organic Photochemistry. Springer-Verlag, Berlin. 1968.
3. Neckers, D. C. Mechanistic Organic Photochemistry. Reinhold, N. Y. 1967.
4. Calvert, J. G. and J. N. Pitts, Jr. Photochemistry, pp.366–685. John Wiley and Sons, N. Y. 1966.
5. Schaffner, K. – Fortsch. Chem. Org. Naturst., 22 : 1. 1964.

6. McLaren, A. D. and D. Shugar. Photochemistry of Proteins and Nucleic Acids. Pergamon Press, N. Y. 1964.
7. Turro, N. J. Molecular Photochemistry, pp. 137–245. W. A. Benjamin, Inc., N. Y. 1965.
8. Chapman, O. L. Advances in Photochemistry. Edited by W. A. Noyes, Jr., G. S. Hammond, and J. N. Pitts, Jr., 1 : 323. Interscience Publishers, N. Y. 1963.
9. Wagner, P. J. and G. S. Hammond. Ibid., 5 : 21. 1968.
10. Reid, S. T. – Recherches Pharmaceutiques. Edited by E. Jucker, **11** : 48. 1968.
11. De Mayo, P. Advances in Organic Chemistry, 2 : 367. Interscience Publishers, N. Y. 1960.
12. Masson, C. R., W. A. Noyes, Jr., and V. Boekelheide. Technique of Organic Chemistry. Edited by A. Weissberger, **2** : 257. Interscience Publishers, N. Y. 1956.
13. Noyes, Jr., W. A. and P. A. Leighton. Photochemistry of Gases. Reinhold, N. Y. 1949.
14. De Mayo, P. and S. T. Reid. – Quart. Rev., **15** : 393. 1961.
15. McNesby, J. R. and H. Okabe. Advances in Photochemistry. Edited by W. A. Noyes, Jr., G. S. Hammond, and J.N.Pitts, Jr., 3 : 157. Interscience Publishers, N. Y. 1964.
16. Srinivasan, R. Ibid., **4** : 113. 1966.
17. Mousseron, M. Ibid., **4** : 195. 1966.
18. Fonken, G. J. Organic Photochemistry. Edited by O. L. Chapman, **1** : 197. M. Dekker, N. Y. 1967.
19. Sternitz, F. R. Ibid., **1** : 247. 1967.
20. Mousseron, M. – Pure Appl. Chem., **9** : 481. 1964.
21. Dauben, W. G. and W. Todd-Pike. – Ibid., **9** : 539. 1964.
22. Prinzbach, H. – Ibid., **16** : 17. 1968.
23. Bartlett, P. D. et al. – Ibid., **16** : 187. 1968.
24. Dauben, W. G. Reactivity of the Photoexcited Organic Molecules, p. 171. Interscience Publishers, N. Y. 1967.
25. Steinmetz, R. – Fortsch. Chem. Forsch., (Photochemie), **7** : 445. 1967.
26. Chapman, O. L. and G. Lenz. Organic Photochemistry. Edited by O. L. Chapman, **1** : 283. M. Dekker, N. Y. 1967.

27. Havinga, E. and M. E. Kronenberg. – Pure Appl. Chem., **16** : 137. 1968.
28. Havinga, E. Reactivity of the Photoexcited Organic Molecules, p. 201. Interscience Publishers, N. Y. 1967.
29. Havinga, E., R. O. de Jongh, and M. E. Kronenberg. – Helv. Chim. Acta, **50** : 2550. 1967.
30. Majer, J. R. and J. P. Simons. Advances in Photochemistry. Edited by W. A. Noyes, Jr., G. S. Hammond, and J.N.Pitts, Jr., **2** : 137. Interscience Publishers, N. Y. 1964.
31. De More, W. B. and S. W. Benson. Ibid., **2** : 219. 1964.
32. Stenberg, V. I. Organic Photochemistry. Edited by O. L. Chapman, **1** : 127. M. Dekker, N. Y. 1967.
33. Barton, D. H. R. – Pure Appl. Chem., **16** : 1. 1968.
34. Elad, D. – Fortsch. Chem. Forsch. (Photochemie), **7** : 528. 1967.
35. Srinivasan, R. – Pure Appl. Chem., **16** : 65. 1968.
36. Weygand, F. and H. J. Bestmann. – Angew. Chem., **72** : 535. 1960.
37. Chapman, O. L. et al. – Pure Appl. Chem., **9** : 585. 1964.
38. Mustafa, A. Advances in Photochemistry. Edited by W. A. Noyes, Jr., G. S. Hammond, and J. N. Pitts, Jr., **2** : 63. Interscience Publishers, N. Y. 1964.
39. Akhtar, M. Ibid., **2** : 263. 1964.
40. Frey, H. M. Ibid., **4** : 225. 1966.
41. Frey, H. M. – Pure Appl. Chem., **9** : 527. 1964.
42. Pape, M. – Fortsch. Chem. Forsch. (Photochemie), **7** : 559. 1967.
43. Müller, E. – Pure Appl. Chem., **16** : 153. 1968.

Chapter 5

PHOTOCHEMISTRY OF BENZENE

I. EXCITED STATES OF BENZENE

1. Symmetry notation of excited states

Up till now we have used the molecular orbital notation – or, more precisely, the simplified notation of Kasha, in which the orbitals not affected by the transition are neglected – to identify the electronic states and transitions, since this notation is the one most frequently employed. Another notation, which is often used, especially for aromatic hydrocarbons, is the symmetry notation. In this notation, a given state is described in terms of the behavior of the electronic wave function in the symmetry operations of the group to which the molecule belongs. Without going into details of group theory, we shall note that each molecule is classified into a symmetry group (D_{6h} for benzene), which comprises different classes (A_1, A_2, B_1, B_2, ..., for benzene). The principal symbol of the state being studied will be the class symbol, e.g., A_1 for the ground state of benzene. The index to the upper left of the principal symbol indicates the multiplicity, while the index to the right and below the principal symbol indicates whether the molecular wave function is symmetrical (g) or antisymmetrical (u) relative to the center of symmetry. The complete notation for the ground state of benzene will be $^1A_{1g}$.

2. Electronic states of benzene

The ground state is $^1A_{1g}$; the first excited singlet state S_1^* is $^1B_{2u}$. The transition $^1B_{2u} \leftarrow {}^1A_{1g}$ involves a change in the symmetry of the molecule and is forbidden. The excited state S_1^* corresponds to a deformation of the geometry of the molecule in the plane.

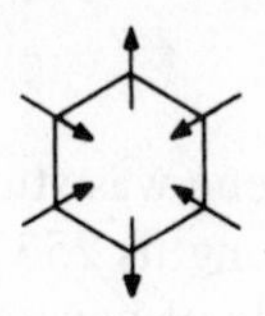

Longuet-Higgins showed that an adiabatic process* may result in a transition from the excited state $^1B_{2u}$ to the biradical form:

$\leftrightarrow$

Thus we may write for the first excited singlet state of benzene:

$$S_0(^1A_{1g}) \xrightarrow{h\nu} S_1^*(^1B_{2u}) \rightarrow \quad \leftrightarrow$$

The biradical form is tautomeric with the state $^1B_{2u}$.

The first triplet state of benzene has also been studied; it is the state $^3B_{1u}$. In this state the molecule is deformed and can be represented as a quinonoid biradical existing in two tautomeric forms:

* In photochemistry (unlike in thermodynamics) the term "adiabatic process" is defined as a process not involving a change in the electronic state.

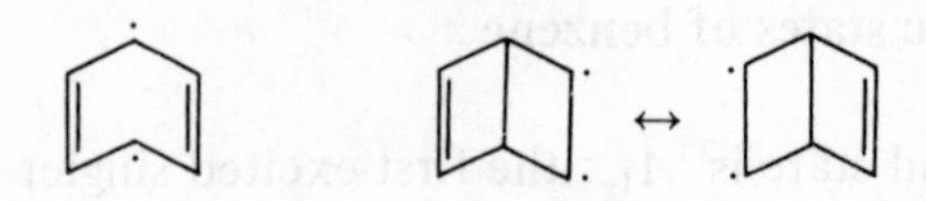

II. PHOTOISOMERIZATION OF BENZENE

1. Irradiation of benzene

The irradiation of benzene was studied both in the liquid and in the vapor phase; the wavelengths 2537 Å or 1849 Å were generally employed. Various isomers of benzene could be isolated and identified, including fulvene (I), benzvalene (II), "Dewar" benzene (III), and prismane (IV).

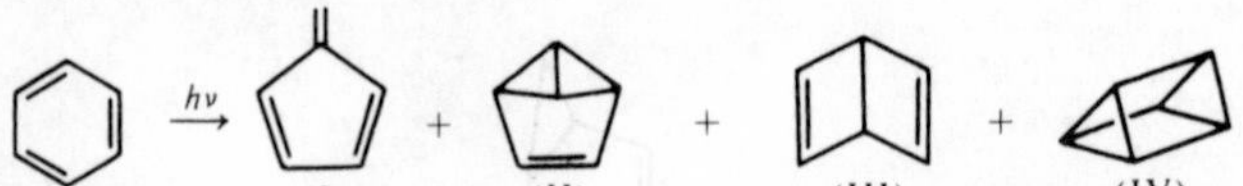

(I) (II) (III) (IV)

According to the mechanism which has been proposed for this isomerization, fulvene (I) and benzvalene (II) originate from the biradical form of the first excited singlet state $^1B_{2u}$:

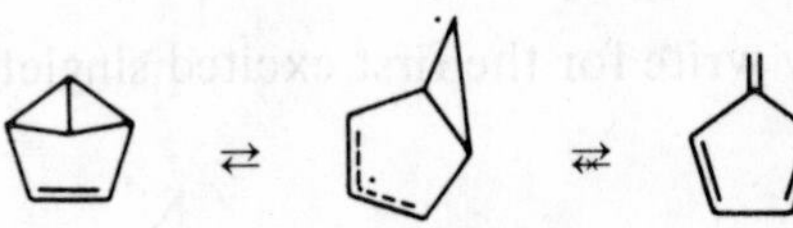

On the other hand, the triplet state $^3B_{1u}$ has been postulated as the intermediate state in the reactions which yield isomers (III) and (IV):

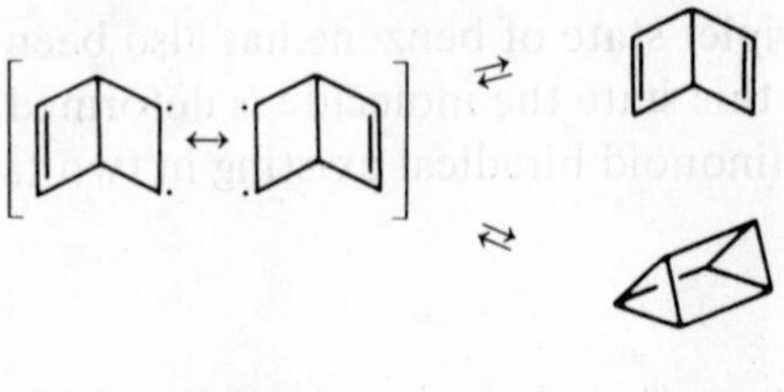

The only isomer which is formed in significant amounts is fulvene since, unlike the other three isomers, it is not reconverted to the aromatic form.

2. Irradiation of benzene in the presence of a monoolefin

In the presence of an olefin R–CH=CH–R′, two types of adducts are formed in lieu of the above isomers:

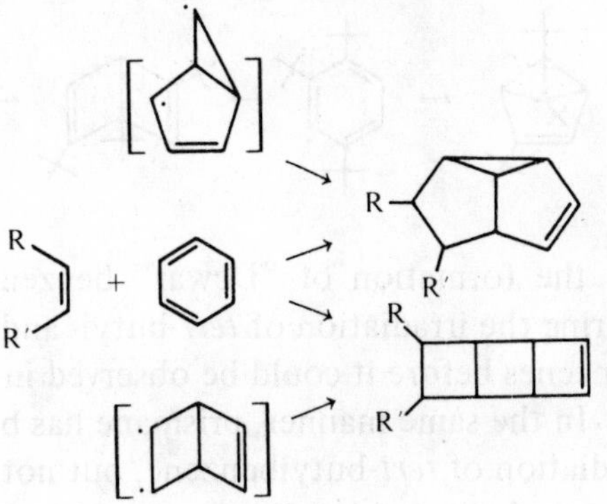

Here, the singlet $^1B_{2u}$ is the parent state of adduct (V), while the triplet $^3B_{1u}$ is the parent state of adduct (VI).

3. Irradiation of benzene in the presence of a 1,3-diene

With 1,3-conjugated dienes, a 1,4-addition takes place on *p*-carbon atoms in the benzene ring. The starting state is believed to be the biradical quinonoid form of the benzene triplet.

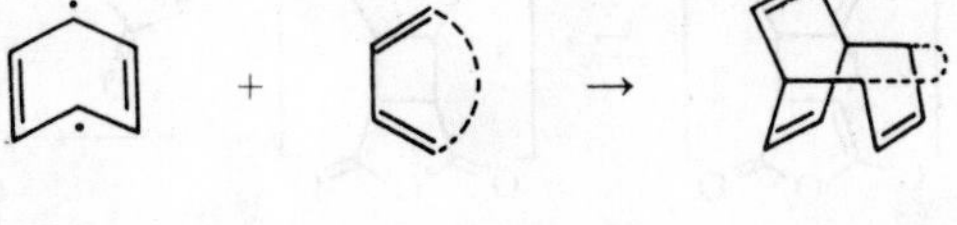

4. Isomerization of substituted benzenes

Since the presence of substituents has a stabilizing effect on the isomeric forms of benzene, these forms could be isolated in larger yields following the irradiation of substituted benzenes.

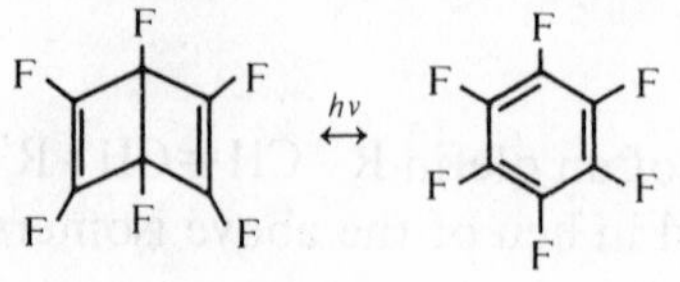

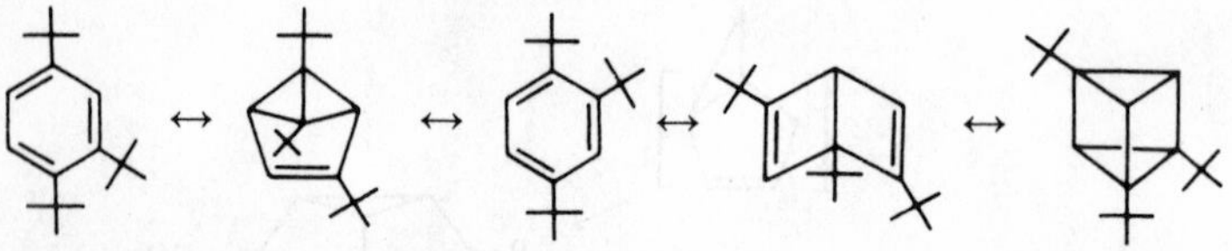

In this way the formation of "Dewar" benzene was demonstrated during the irradiation of *tert*-butyl- and fluoro-substituted benzenes before it could be observed in the case of benzene itself. In the same manner, prismane has been isolated during the irradiation of *tert*-butylbenzene, but not during the irradiation of nonsubstituted benzene.

III. PHOTOCHEMICAL ADDITION OF MALEIC ANHYDRIDE TO BENZENE

Several workers have observed that an adduct of benzene and maleic anhydride is formed in excellent yields; its structure has been determined as (VIII).

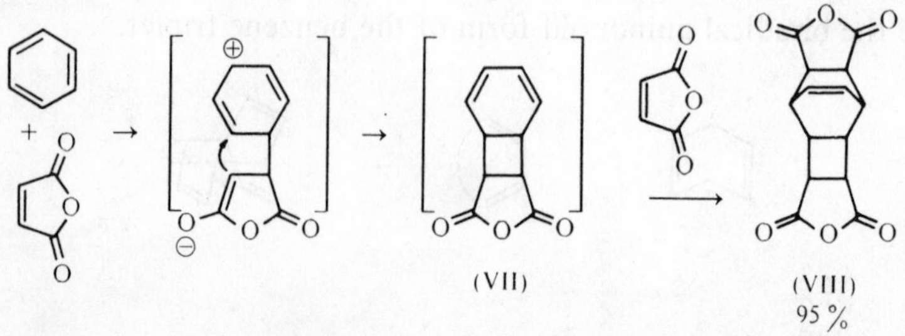

(VII)

(VIII)
95 %

Since benzene and maleic anhydride are capable of forming a charge transfer complex, a mechanism involving an excited state of this complex was postulated. This mechanism has been verified: the addition reaction does not proceed in cyclohexane, since a charge transfer complex is not formed in this solvent. The adduct (VIII) originates from a Diels-Alder addition of a molecule of maleic anhydride to the intermediate complex (VII) (which cannot be isolated), formed by the transition of the benzene-maleic anhydride charge transfer complex to the excited state. It appears that the reaction proceeds via the excited singlet state of the complex in the absence of a sensitizer, and via the triplet state if the reaction is photosensitized. The energy of the triplet state of the complex has been estimated at about 65 kcal$\cdot$mole^{-1}.

IV. PHOTOOXIDATION OF BENZENE

If benzene is irradiated by X-rays, γ-rays, or β-rays in the presence of water and oxygen, a mixture of phenols and muconic aldehyde is obtained; it has been suggested that the reaction involves an attack on benzene by an OH$\cdot$ radical. If the UV 2537 Å radiation is made to act on aqueous benzene at an acidic pH, phenolic products only are formed, either in the presence or in the absence of oxygen; if the pH is neutral, on the other hand, muconic aldehyde is the main product in the presence of oxygen. In such a case, the phenolic product probably consists of hydroquinone, or pyrocatechol, or a mixture of both, rather than of phenol.

According to a recent study, irradiation of benzene in the presence of oxygen but in the absence of water with 2537 Å light results in the formation of *trans-trans*-muconic aldehyde (IX) and of 2, 4, 6, 8, 10-dodecapentaenedial (X). The formation of phenolic compounds takes place even in the absence of water. It is believed that aldehydes (IX) and (X) are formed from anhydrous benzene via the triplet state of benzene:

$$\xrightarrow{h\nu} {}^1B_{2u} \rightsquigarrow {}^3B_{1u} \leftrightarrow \xrightarrow{O_2}$$

$$\rightarrow \left[\begin{matrix} CHO \\ CHO \end{matrix} \right] \rightarrow OHC\text{—}CH{=}CH\text{—}CH{=}CH\text{—}CHO$$

(IX)

(X)

A study of the reaction kinetics (quenching) has confirmed that the reaction mechanism in fact involves the state ${}^3B_{1u}$ of benzene. The kinetics results were confirmed by formation of dialdehyde by direct excitation* ${}^1A_{1g} \rightarrow {}^3B_{1u}$ at 3130Å.

V. OTHER PHOTOCHEMICAL REACTIONS OF BENZENE

Photooxidation is not the only example of a photochemical rupture of the aromatic ring. Irradiation of benzene in a rigid medium yields a hexatrienyl radical (XI) which usually adds on to the solvent.

$$\xrightarrow{h\nu} \cdot CH{=}CH{-}CH{=}CH{-}CH{=}CH\cdot$$

(XI)

* In the presence of oxygen, the intensity of the forbidden absorption $S_0 \rightarrow T_1$ of benzene is greatly increased. Mulliken explains this effect by postulating a benzene-oxygen charge transfer complex. Hence, it is possible to excite the benzene triplet directly and thus to eliminate the reactivity of the singlet state.

Here, too, the triplet state $^3B_{1u}$ is supposed to be the precursor of the reaction.

The addition of benzene to certain acetylenic compounds eventually yields a cyclooctatetraene. The intermediate compound may be the product of an initial 1,2-addition, comparable to the addition of monoolefins (see Section II, 2), with or without isomerization of the benzene ring:

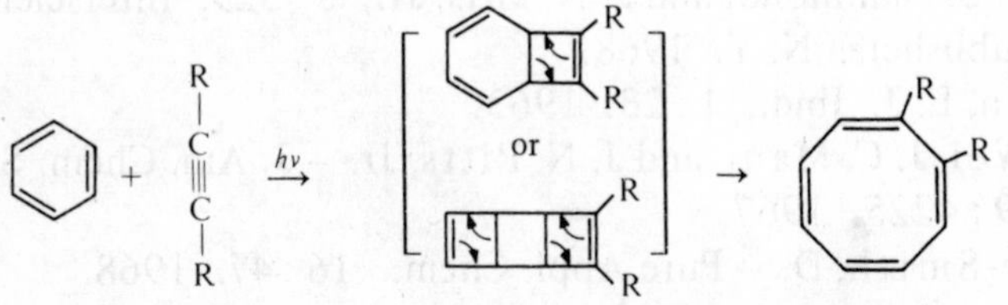

Other reactions given by benzene include dimerization to biphenyl and polymerization, which often appears as a parasitic side reaction. Photochemical 1,4-addition of pyrrole to benzene was recently observed:

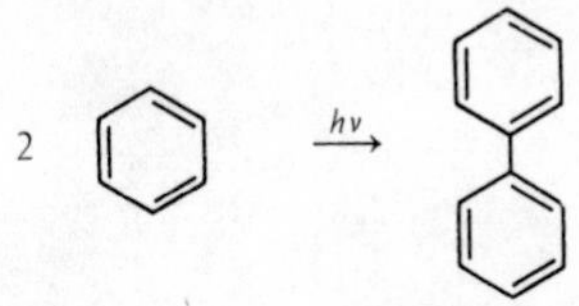

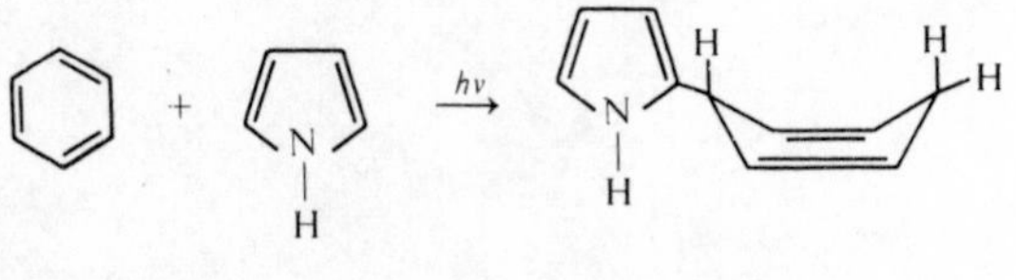

Bibliography

1. Leach, S. and R. Lopez-Delgado. – J. Chim. Phys., p.1636. 1964.

2. Angus, H. J. F., J. McDonald Blair, and D. Bryce-Smith. – J. Chem. Soc., 2003. 1960.
3. Bryce-Smith, D. and H. C. Longuet-Higgins. – Chem. Comm., **17**: 593. 1966.
4. Tsubomura, H. and R. S. Mulliken. – J. Am. Chem. Soc., **82**: 5966. 1960.
5. Phillips, D., J. Lemaire, C. S. Burton, and W. A. Noyes, Jr. Advances in Photochemistry. Edited by W. A. Noyes, Jr., G. S. Hammond, and J. N. Pitts, Jr., **5**: 329. Interscience Publishers, N. Y. 1968.
6. Bowen, E. J. Ibid., **1**: 23. 1963.
7. Kei Wei, J. C. Mani, and J. N. Pitts, Jr. – J. Am. Chem. Soc., **89**: 4225. 1967.
8. Bryce-Smith, D. – Pure Appl. Chem., **16**: 47. 1968.

Chapter 6

ORBITAL SYMMETRY AND CONCERTED REACTIONS. WOODWARD-HOFFMANN RULES

Recent studies, mainly those of Woodward and Hoffmann, Longuet-Higgins and Abrahamson, Oosterhoff, and Havinga, showed that the mechanism and the stereochemistry of certain concerted reactions can be predicted on the strength of the symmetry features of electronic states. Orbital symmetry determines the feasibility and the stereochemical course of each concerted reaction.

I. ORBITAL SYMMETRY OF POLYENIC SYSTEMS

The method of linear combination of atomic orbitals (LCAO), which was applied by Hueckel to polyenic systems, gives an approximate expression for a molecular orbital resulting from a linear combination of atomic orbitals. For a polyenic chain having n carbon atoms and n orbitals π, $\varphi_1, \ldots, \varphi_n$, each of them on one carbon atom, it is possible to construct n molecular orbitals $\psi_1, \ldots, \psi_n$:

$$\psi_j = \sum_{i=1}^{i=n} a_{ij} \cdot \varphi_i$$

where a_{ij} are numerical coefficients which describe the relative contributions of the different atomic orbitals. For the above polyene:

$$a_{ij} = \sqrt{\frac{2}{n+1}} \cdot \sin\left(ij \cdot \frac{\pi}{n+1}\right).$$

We distinguish between two groups of molecular orbitals: symmetrical orbitals S for which:

$$a_{1,j} = a_{n,j}\,;\quad a_{2,j} = a_{n-1,j}\,;\quad a_{3,j} = a_{n-2,j} \ldots \text{etc.}\,;$$

and antisymmetrical orbitals A for which

$$a_{1,j'} = -\,a_{n,j'}\,;\quad a_{2,j'} = -\,a_{n-1,j'} \ldots \text{etc.}\,.$$

It is important to note that the sequence of the orbitals is always S, A, S, A ..., the lowest orbital being always S.

The simplest polyene – butadiene – will have four π-orbitals, viz. ψ_1, ψ_2, ψ_3, and ψ_4, this being the sequence of increasing energies (Figure 6.1).

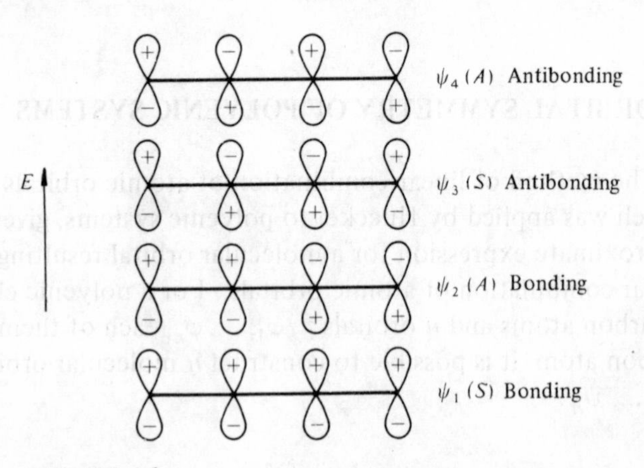

FIGURE 6.1

In an *S*-orbital, which can be bonding or antibonding, the signs of the wave function on the two end carbons in the chain are the same. In an *A*-orbital, which can also be bonding or antibonding, these signs are opposite.

The highest occupied molecular orbital will determine the course of the reaction. It is accordingly essential to determine the symmetry of this orbital, in the ground state and in the first excited state, as a function of the number of π-electrons of the system.

The ground state is obtained by placing pairs of electrons in the molecular orbitals, in the sequence of increasing energy. The highest occupied molecular orbital will be *S* or *A*, depending on whether the number of occupied orbitals is odd or even, respectively. In the case of butadiene, in which the four π-electrons occupy the orbitals ψ_1 and ψ_2 in the ground state, the highest occupied orbital ψ_2 is in fact of the *A* type.

The first excited state corresponds to a passage of an electron from the highest occupied to the lowest vacant orbital; in this state, the energetically highest occupied orbital will be *S* if the highest occupied orbital was *A* in the ground state and vice versa.

The number of molecular π-orbitals occupied in the ground state can obviously be deduced from the number of π-electrons of the molecule. If the number of π-electrons in the molecule is $n = 2p$, the number of occupied π-orbitals in the ground state will be p; for butadiene $n = 4$ and $p = 2$.

The same reasoning applies to a free radical or to an ion. In the case of an **allyl radical** $CH_2{=}CH{-}CH_2\cdot$, there are three π-electrons, and two occupied orbitals in the ground state; thus the highest occupied orbital, with only one electron in it, will be *A*. In the first excited state, the highest occupied orbital will be *S*.

An **allyl cation** $CH_2{=}CH{-}CH_2^+$ has two π-electrons and only one orbital (an *S*-orbital) occupied in the ground state. In the first excited state, the highest occupied orbital will be *A*.

An **allyl anion** $CH_2{=}CH{-}CH_2{:}^-$ has four π-electrons and two occupied orbitals in the ground state, so that the highest occupied orbital is an *A*-orbital. In the first excited state, the highest occupied orbital is *S*.

In the general case, a cation or an anion will behave similarly to a molecule with the same number of π-electrons; a free radical will behave like a molecule with one additional π-electron. This rule has been confirmed for radicals in the ground state; in the excited state, the problem is more difficult and will be discussed below.

II. ELECTROCYCLIC TRANSFORMATIONS

This kind of transformation has been defined by Woodward as the formation or the rupture of a σ-bond between the two ends of a polyenic chain. These reactions are highly stereospecific and their stereospecificity depends both on the identity of the polyene and on the nature of the activation process (thermal or photochemical). Orbital symmetry can account for this stereospecificity in the following manner.

During an "allowed" concerted reaction (i.e., "allowed" from the viewpoint of symmetry), in which a molecular system X (comprising one or more molecules) is transformed into a molecular system Y, it is possible to effect a continuous transition from the molecular orbitals of X to those of Y so that the bonding nature of all the occupied molecular orbitals is preserved throughout the course of the reaction.

A σ-bond between the two ends of a polyenic chain is formed as a result of the overlap of two lobes of the same sign of the π-orbitals of the two chain-end carbons, in the highest occupied molecular orbital. If the wave functions on the two terminal carbon atoms have the same sign (S orbital), the ring closure will be "disrotatory."* If the signs are different, the ring closure will be "conrotatory" (Figure 6.2).

* Two different conrotatory and two different disrotatory movements can be distinguished, depending on whether it is the (+) lobes or the (–) lobes of the orbitals which are combined; the stereochemistry may be different in the two cases. The Woodward-Hoffmann rules do not make these distinctions. The movements can be differentiated with the aid of other (most often steric) considerations (e.g., opening of cyclopropyl cation ring, p.111).

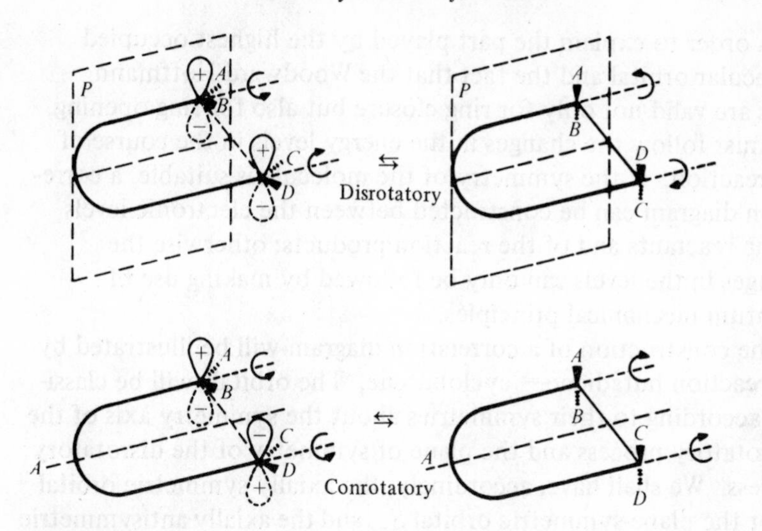

FIGURE 6.2

The disrotatory process has a plane of symmetry which is perpendicular to the plane of the polyenic chain. The conrotatory process has an axis of symmetry in the plane of the chain.

The highest occupied orbital will be *A* or *S* depending on the number of occupied π-orbitals in the ground state; accordingly, the nature of the concerted cyclization process will depend on the number of the π-electrons in the system. Also, since the highest occupied orbital has a different symmetry for the ground state and for the first excited state, thermal ring closure (ground state) will differ from photochemical ring closure (first excited state). Accordingly, the knowledge of the number of π-electrons in the system makes it possible to predict the mode of thermal or photochemical ring closure (Woodward-Hoffmann rules, Table 6.1).

TABLE 6.1. The Woodward-Hoffmann rules of electrocyclic transformations

Number of π-electrons	Thermal	Photochemical
$4q$	Conrotatory	Disrotatory
$4q + 2$	Disrotatory	Conrotatory

In order to explain the part played by the highest occupied molecular orbital and the fact that the Woodward-Hoffmann rules are valid not only for ring closure but also for ring opening, we must follow the changes in the energy levels in the course of the reaction. If the symmetry of the molecule is suitable, a correlation diagram can be constructed between the electronic levels of the reactants and of the reaction products; otherwise the changes in the levels can only be followed by making use of quantum mechanical principles.

The construction of a correlation diagram will be illustrated by the reaction butadiene ⇌ cyclobutene. The orbitals will be classified according to their symmetries about the symmetry axis of the conrotatory process and the plane of symmetry of the disrotatory process. We shall have, accordingly, the axially symmetric orbital S_A or the plane-symmetric orbital S_p, and the axially antisymmetric orbital A_A or the plane-antisymmetric orbital A_p.

We shall consider the opening of the cyclobutene ring (Figure 6.3):

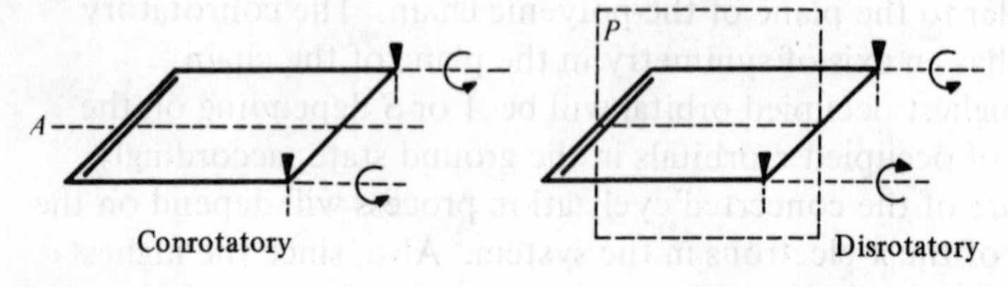

FIGURE 6.3

We have to consider the four molecular orbitals of butadiene π_1, π_2, π_1^*, π_2^*, and the four molecular orbitals of cyclobutene σ, π, π^*, σ^*, and arrange them in an increasing energy sequence (Figure 6.4).

Since only two orbitals with the same symmetry type are compatible, we can establish the following correlation diagrams* (Figure 6.5).

* The crossing of correlation lines of the same symmetry is forbidden:

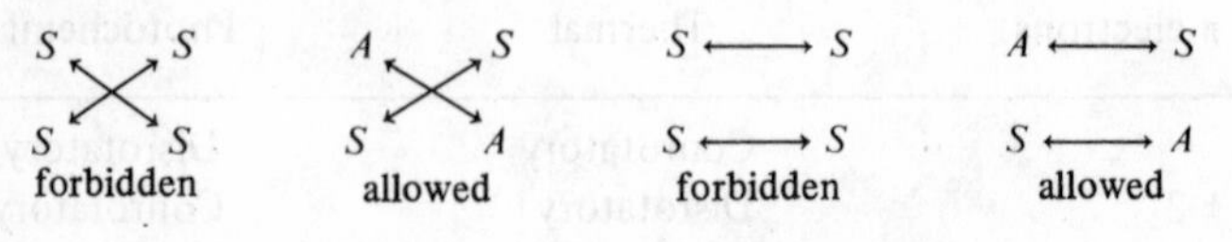

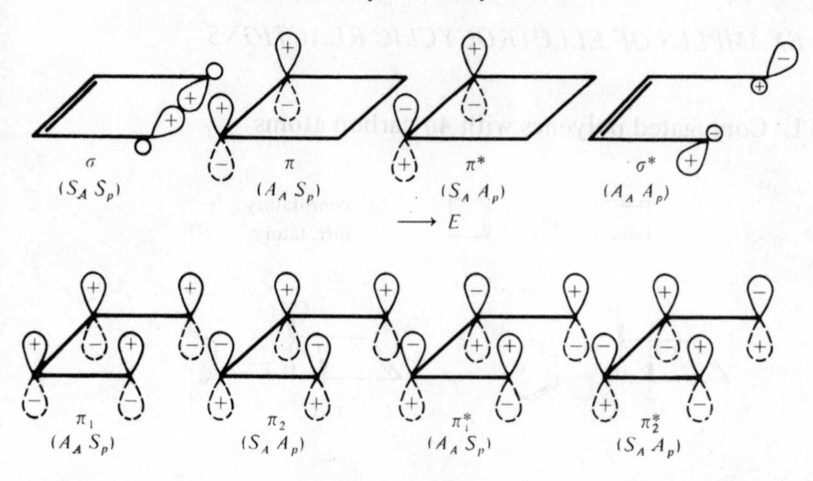

FIGURE 6.4

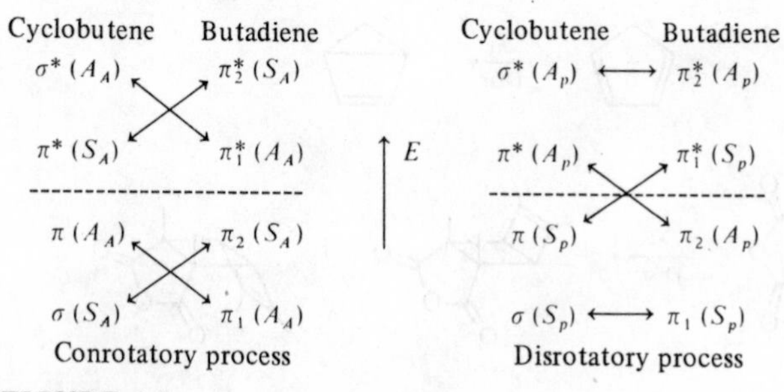

FIGURE 6.5

In the ground state, cyclobutene has two electrons each in a σ-orbital and a π-orbital, while butadiene has two electrons each in a π_1-orbital and a π_2-orbital. It is clear from the above diagram that the reaction cyclobutene $\rightleftharpoons$ butene can be effected by a conrotatory process while remaining in the ground state (thermal reaction), whereas the disrotatory process involves an antibonding orbital (photochemical reaction).*

* Another way of saying the same thing is that the reaction is "allowed" if it is accompanied by a decrease in the energy of the highest occupied orbital; otherwise, it is "forbidden."

EXAMPLES OF ELECTROCYCLIC REACTIONS

1. Conjugated polyenes with 4*q* carbon atoms

Δ : conrotatory
$h\nu$: disrotatory

CH_3 CH_3 H H 200° Conr.

CH_3 H H CH_3 200° Conr.

Δ Conr.

$h\nu$ Disr.

$h\nu$ Disr.

O O O $h\nu$ Disr. O O O + O O O

$h\nu$ Disr.

Certain cyclic polyenes with $4q + 2$ carbon atoms behave as polyenes with $4q$ carbon atoms, with only a part of the polyenic system participating in the electrocyclic reaction.

2. Conjugated polyenes with $4q + 2$ carbon atoms

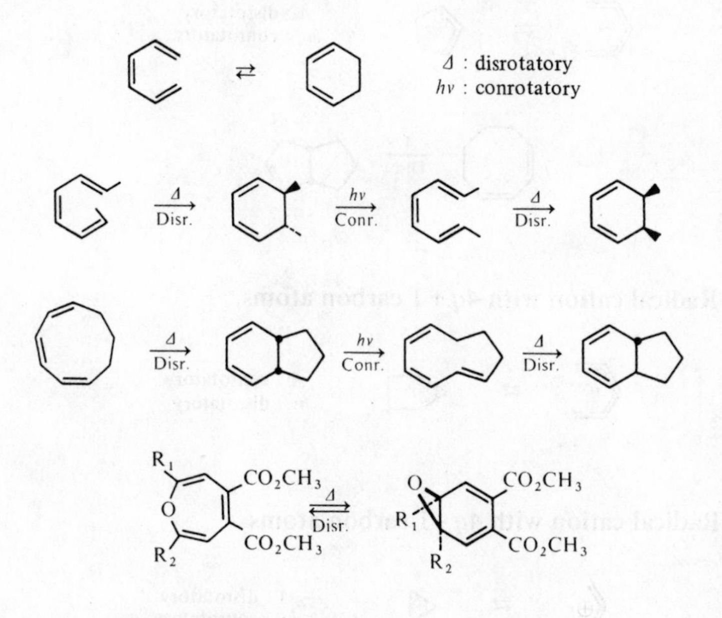

3. Radical anion with $4q - 1$ carbon atoms

Δ : conrotatory
$h\nu$: disrotatory

Thermal ring closure of a heptatrienyl anion originating from octatriene was recently observed:

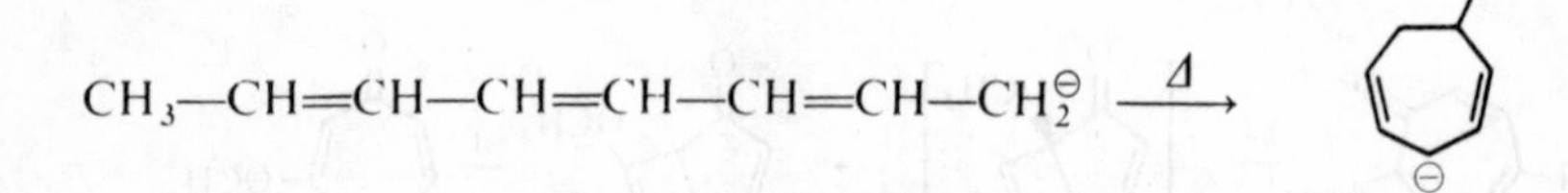

4. Radical anion with $4q + 1$ carbon atoms

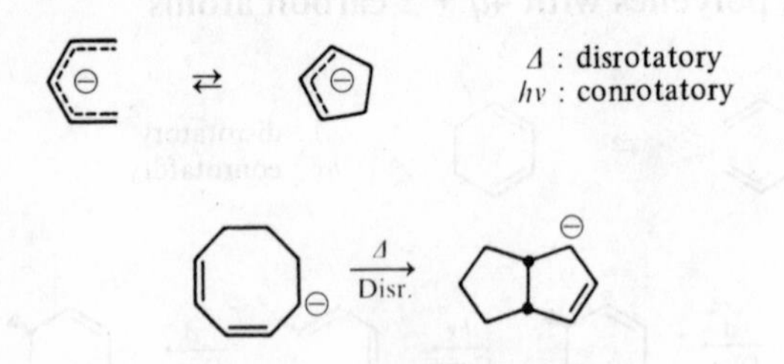

5. Radical cation with $4q + 1$ carbon atoms

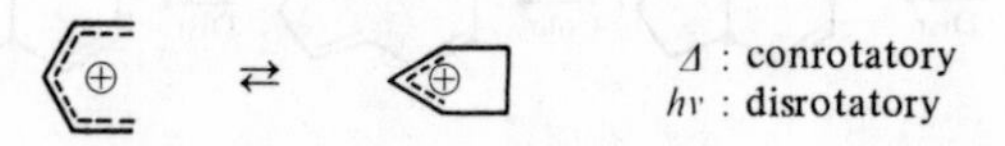

6. Radical cation with $4q - 1$ carbon atoms

Δ : disrotatory
$h\nu$: conrotatory

In the cyclopropane compounds below, the elimination of substituent X (halogen or tosylate) with formation of a cyclopropyl cation is concerted with ring opening to give an allyl cation. The disrotatory process can proceed in two ways, but in each case process 1 will be favored (Figure 6.6).

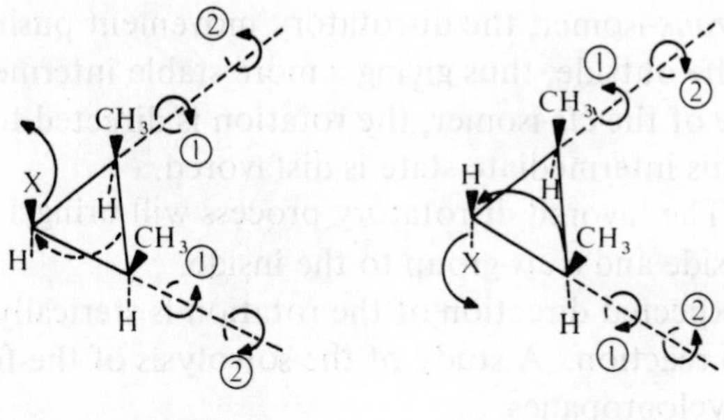

FIGURE 6.6

Process 1 facilitates the elimination of X, since it involves the overlap of the rear lobe of the C–X orbital by the 2*p* orbitals formed on the terminal carbons as a result of the opening of the σ-bond (Figure 6.7).

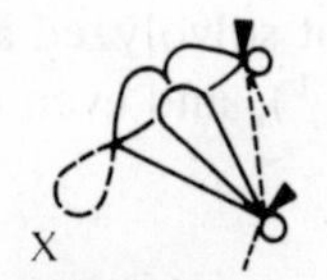

FIGURE 6.7

Thus, *trans*-arylcyclopropyl tosylates are solvolyzed at more than 15 times the rate of their *cis*-isomers:

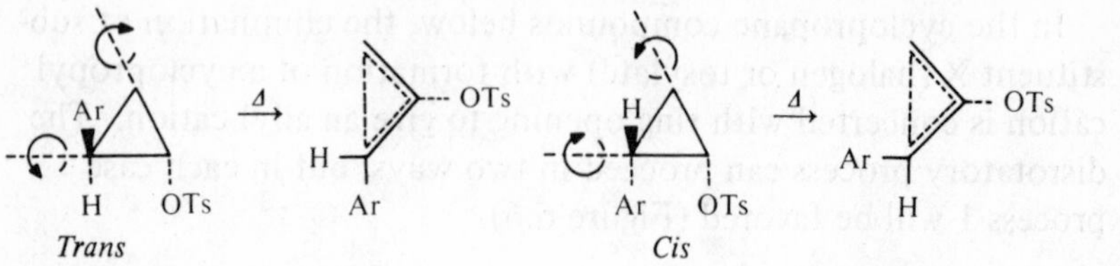

Ar OTs H Ar $\xrightarrow{\Delta}$ H Ar OTs   H Ar OTs $\xrightarrow{\Delta}$ Ar H OTs

Trans   *Cis*

In the *trans*-isomer, the disrotatory movement pushes the phenyl group to the outside, thus giving a more stable intermediate state. In the case of the *cis*-isomer, the rotation is directed toward the interior; this intermediate state is disfavored.

Rule: The favored disrotatory process will bring a *trans*-group to the outside and a *cis*-group to the inside.

If the expected direction of the rotation is sterically impossible, there is no reaction. A study of the solvolysis of the following two isomeric cyclopropanes

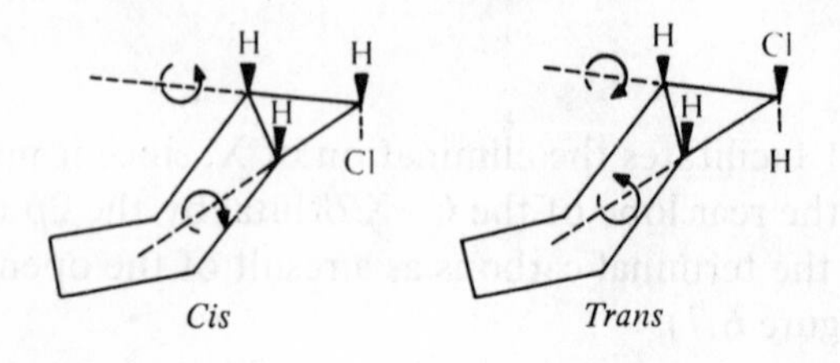

Cis   *Trans*

showed that the *cis*-isomer is solvolyzed at 125° ($k = 1.4 \cdot 10^{-6}\,\text{sec}^{-1}$), while the *trans*-isomer is not solvolyzed at this temperature even in 692 hours ($k < 8 \cdot 10^{-9}\,\text{sec}^{-1}$), and even not at 210°C.

7. Free radicals

It should be noted that caution must be exercized in applying the Woodward-Hoffmann rules to free radicals. Thus, for instance, the first excited state of the allyl radical $CH_2{=}CH{-}CH_2\cdot$ is represented by two quasi-degenerate configurations, which are equivalent (Figure 6.8).

The conrotatory process is thermally favored. The question is still open as regards the photochemical reaction.

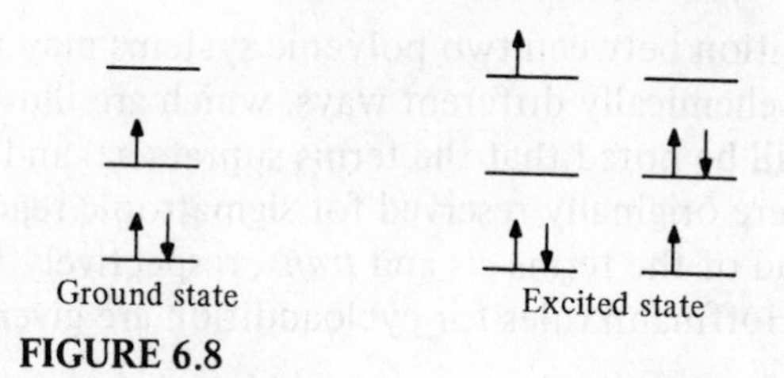

FIGURE 6.8

III. CONCERTED CYCLOADDITIONS

We shall discuss the concerted additions of the termini of two polyenic systems with m and n π-electrons respectively, and the cycloadditions of more than two systems.

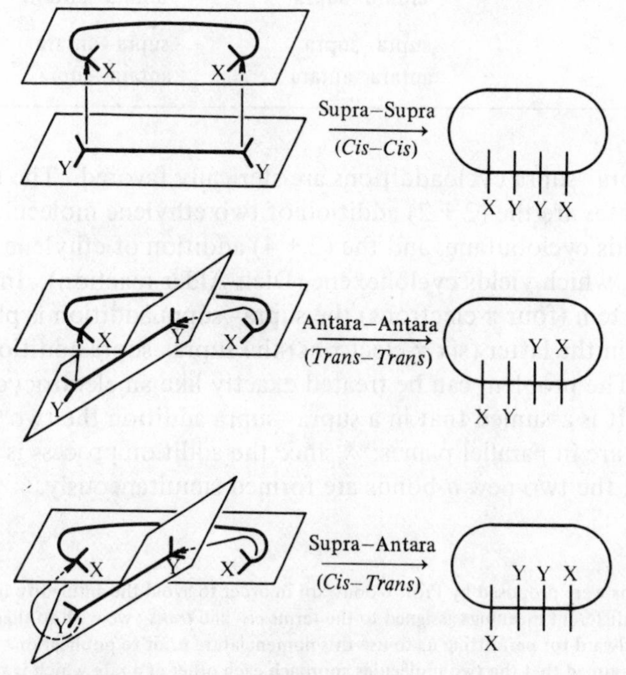

FIGURE 6.9. Supra–antara and antara–supra processes can be stereochemically different. It should also be noted that antara–antara and supra–antara (or antara–supra) processes are in general sterically disfavored; supra–supra reactions are the most common.

A concerted addition between two polyenic systems may take place in four stereochemically different ways, which are illustrated in Figure 6.9. It will be noted that the terms suprafacial and antarafacial, which were originally reserved for sigmatropic reactions, are here used instead of the terms *cis* and *trans*, respectively.*

The Woodward-Hoffmann rules for cycloaddition are given in Table 6.2.

TABLE 6.2

$m + n$	Thermal	Photochemical
$4q$	supra–antara antara–supra	supra–supra antara–antara
$4q + 2$	supra–supra antara–antara	supra–antara antara–supra

The supra–supra cycloadditions are sterically favored. The two classical cases are the (2 + 2) addition of two ethylene molecules, which yields cyclobutane, and the (2 + 4) addition of ethylene to butadiene, which yields cyclohexene (Diels-Alder reaction). In the former system (four π-electrons) the supra–supra addition is photochemical; in the latter (six π-electrons) the supra–supra addition is thermal. The problem can be treated exactly like an electrocyclic reaction. It is assumed that in a supra–supra addition the two molecules are in parallel planes;** since the addition process is concerted, the two new σ-bonds are formed simultaneously.

* These terms were proposed by Prof. Woodward in order to avoid the ambiguity inherent in the different meanings assigned to the terms *cis-* and *trans-*; we wish to thank Prof. Woodward for permitting us to use this nomenclature prior to publication.

** It is also assumed that the two molecules approach each other at a rate which is sufficiently slow for the perturbation theory to be applicable.

1. (2 + 2) Supra–supra addition

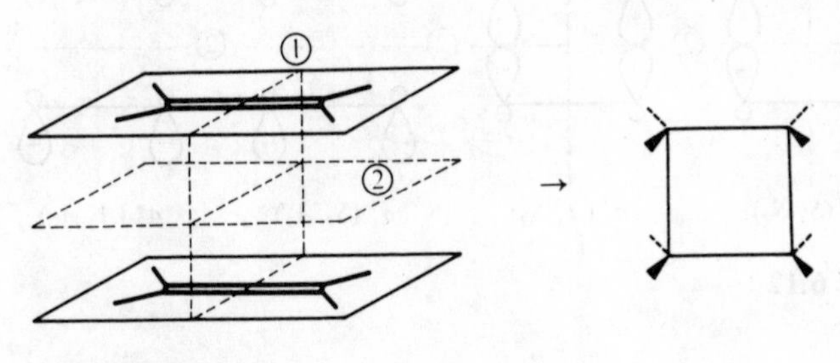

FIGURE 6.10

A system of two ethylene molecules has two planes of symmetry: plane 2, which is parallel to the planes of the molecules, and plane 1, which is perpendicular to the planes of the molecules and which cuts across the two double bonds. The orbitals of the system of the two ethylene molecules are, in increasing energy sequence, π_1, π_2, π_1^*, π_2^* (Figure 6.11):

π_1 is the bonding combination of the π-orbitals of the two ethylenes, which corresponds to a decrease in energy;

π_2 is the antibonding combination of the π-orbitals of the two ethylenes, which corresponds to an increase in energy;

in the same manner, π_1^* and π_2^* are, respectively, the bonding and the antibonding combinations of the π^*-orbitals of the ethylenes.

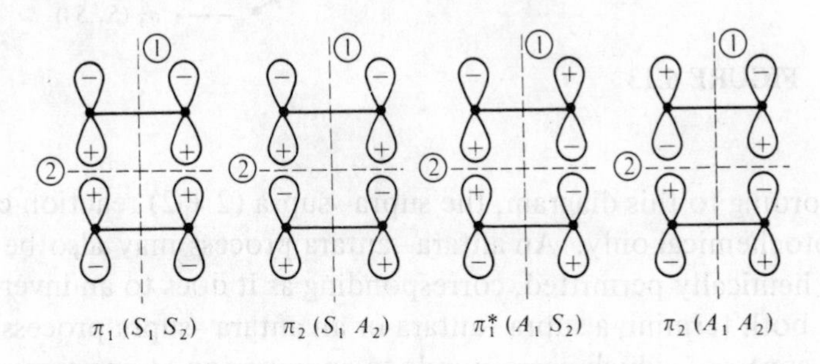

FIGURE 6.11

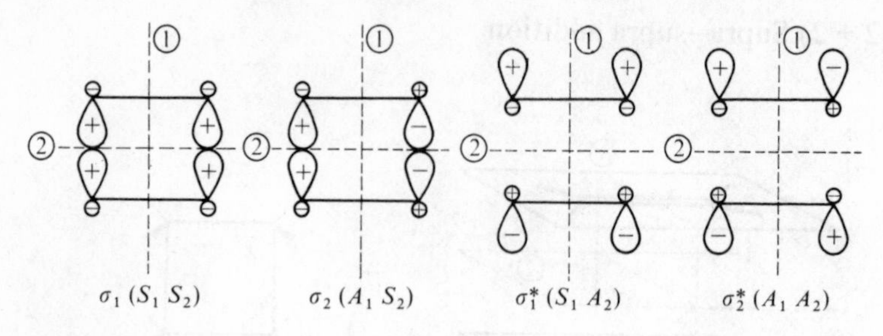

FIGURE 6.12

The cyclobutane product has four new molecular sigma orbitals: σ_1, σ_2, σ_1^*, σ_2^* (Figure 6.12).

The correlation diagram is shown in Figure 6.13:

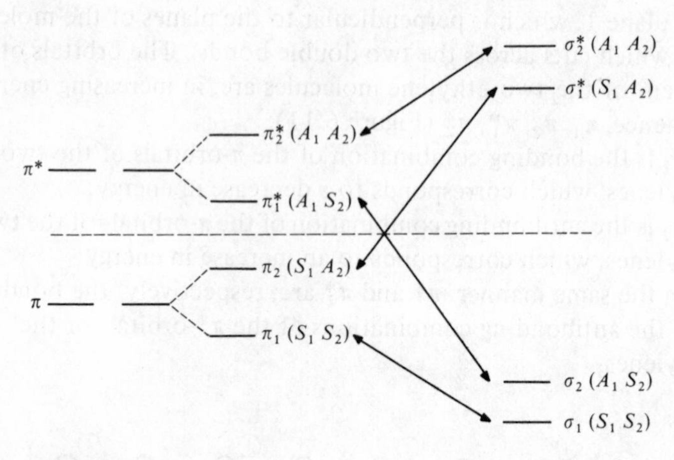

FIGURE 6.13

According to this diagram, the supra–supra (2 + 2) reaction can be photochemical only. An antara–antara process may also be photochemically permitted, corresponding as it does to an inversion at both termini; a supra–antara or an antara–supra process, on the contrary, which corresponds to an inversion at one terminus only, i.e., to an inverse symmetry, can be thermal only.

2. (4 + 2) Supra–supra addition

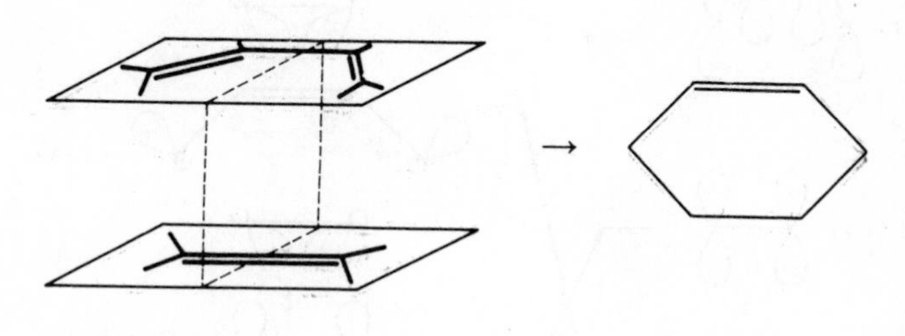

FIGURE 6.14

The system has a single plane of symmetry, and the orbitals are classified with respect to this plane (Figures 6.14 and 6.15).

This diagram shows that there is no correspondence between the occupied and the vacant molecular orbitals; a discussion similar to that given for (2 + 2) addition will show that the antara–antara process may be thermal or photochemical, whereas the supra–antara and the antara–supra processes are exclusively photochemical.

3. Secondary interactions between orbitals

Let us consider, as an example, the Diels-Alder reaction between two molecules of butadiene; the supra–supra approach between the two molecules can be effected in two ways: endo and exo (Figure 6.16).

In order to find out which is the favored approach, we must recall, first and foremost, that a secondary interaction between two occupied orbitals of the diene and the dienophile would have only a negligible effect, since it would stabilize one orbital and destabilize the other. The major interactions will originate from mixing of a filled with an empty level: the highest occupied

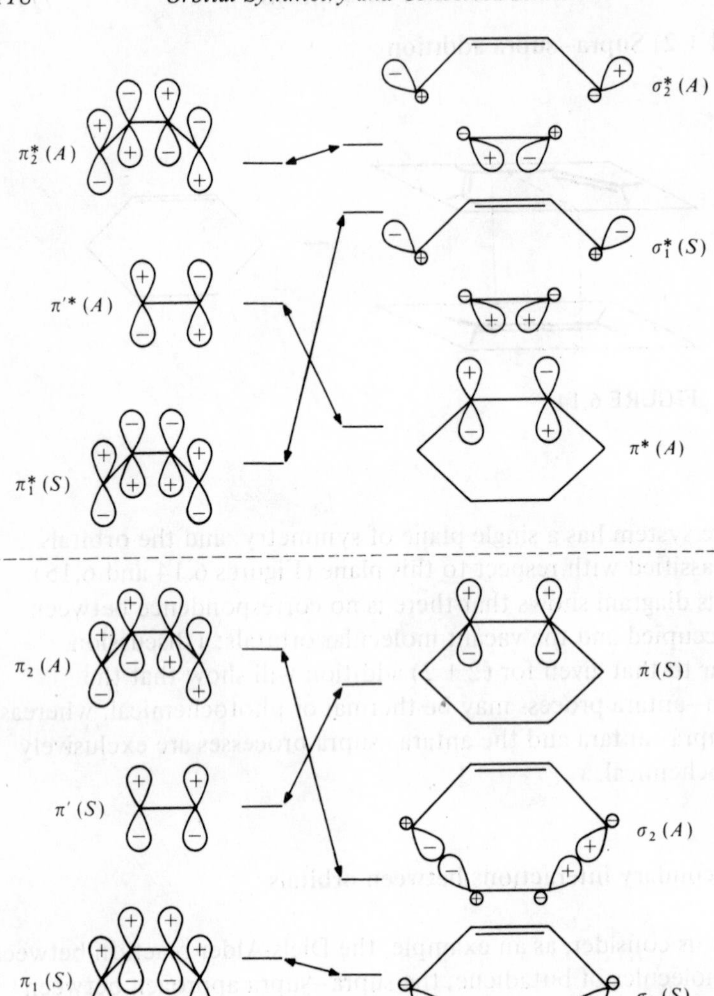

σ_2^* (A)
π_2^* (A)
σ_1^* (S)
π'^* (A)
π^* (A)
π_1^* (S)
π_2 (A)
π (S)
π' (S)
σ_2 (A)
π_1 (S)
σ_1 (S)

FIGURE 6.15

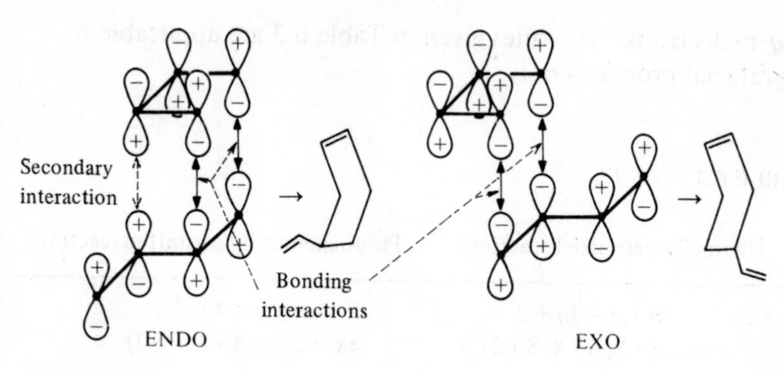

FIGURE 6.16

orbital of the diene and the lowest vacant orbital of the dienophile or vice versa. The end result is the same: it is seen that in this Diels-Alder reaction the endo-addition is favored by secondary orbital interaction (because a secondary bonding interaction reduces the total energy of the system); this is what is in fact observed. This mixing of the highest occupied orbital with the lowest vacant orbital is symmetrically allowed. The stabilization by these secondary interactions will be more considerable, the smaller the energy difference between the two interacting orbitals.

Secondary interactions are involved in all concerted cycloadditions. It may be predicted, for example, that in the (6 + 4) addition of hexatriene to butadiene, the exo process will be favored. This was recently confirmed by Cookson (p.124).

It should be borne in mind that secondary orbital interactions are accompanied by small energy effects and may be compensated by other low-energy effects such as the effect of the substituent, of the solvent, etc.

4. General rules of concerted cycloaddition

The Woodward-Hoffmann rules also apply to systems with a more involved structure: formation of 3 or 4 σ-bonds from 3 or 4 π-bonds by the addition of 3 or 4 polyenic systems with m, n,

p, q π-electrons. The rules given in Table 6.3 are applicable to suprafacial processes only.

TABLE 6.3

	Thermally permitted reactions	Photochemically permitted reactions
$2\pi \rightarrow 2\sigma$	$m + n = 4q + 2$ (ex.: 4+2, 6+4, 8+2)	$m + n = 4q$ (ex.: 2+2, 4+4, 6+2)
$3\pi \rightarrow 3\sigma$	$m + n + p = 4q + 2$ (ex.: 2 + 2 + 2, 2 + 4 + 4, 6 + 2 + 2)	$m + n + p = 4q$ (ex.: 4 + 2 + 2)
$4\pi \rightarrow 4\sigma$	$m + n + p + q = 4q + 2$ (ex.: 4 + 2 + 2 + 2)	$m + n + p + q = 4q$ (ex.: 2 + 2 + 2 + 2)

5. Examples of cycloaddition

(2 + 2) *addition*

Supra – supra photochemical addition is very frequently observed:

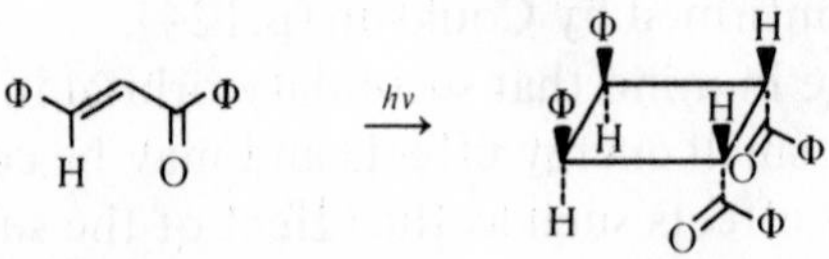

Φ H O Φ $\xrightarrow{h\nu}$ Φ H Φ H H Φ O H O Φ

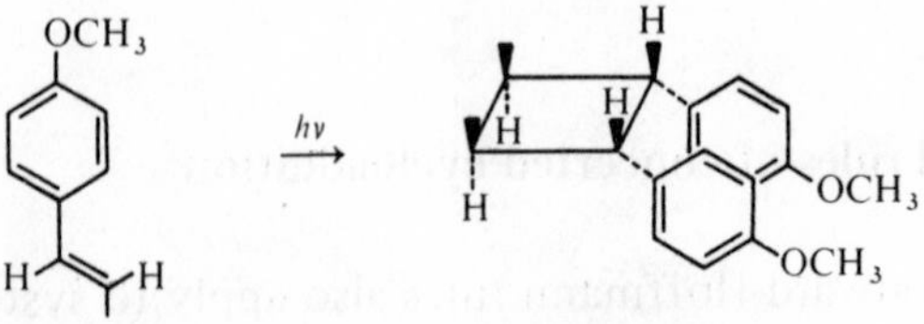

OCH_3 H H $\xrightarrow{h\nu}$ H H H H OCH_3 OCH_3

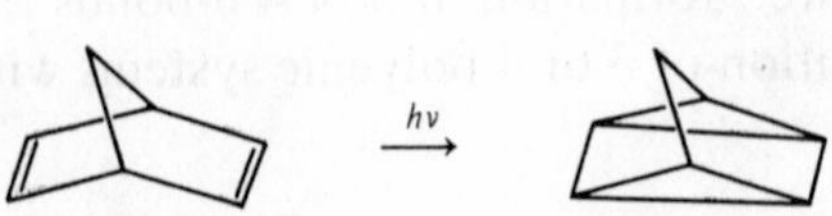

$\xrightarrow{h\nu}$

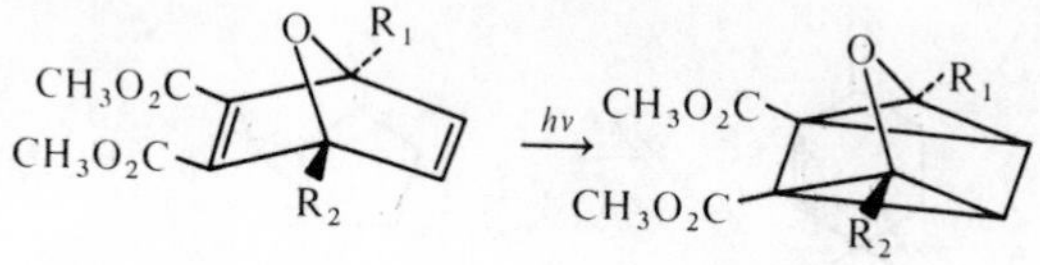

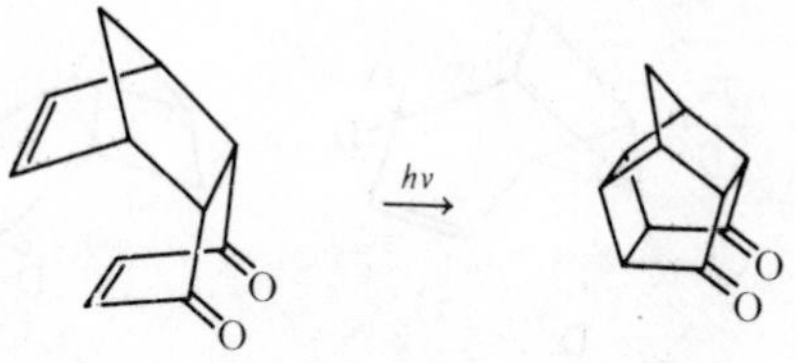

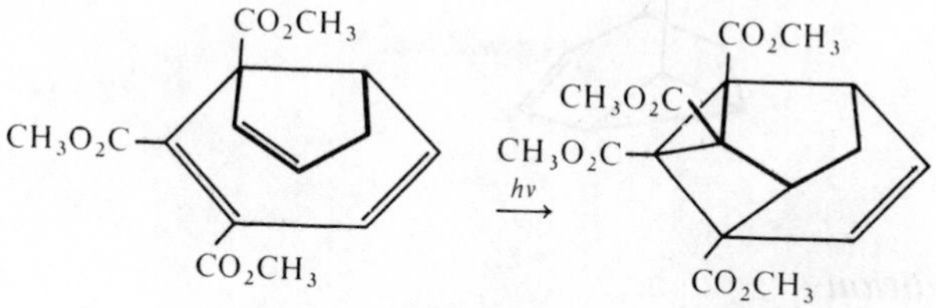

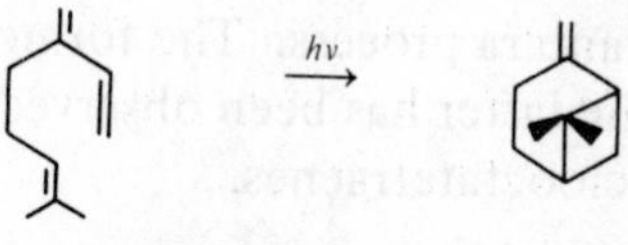

CH_3O_2C CH_3O_2C R_1 R_2 $\xrightarrow{h\nu}$ CH_3O_2C CH_3O_2C R_1 R_2

$\xrightarrow{h\nu}$

CO_2CH_3 CH_3O_2C CO_2CH_3 $\xrightarrow{h\nu}$ CO_2CH_3 CH_3O_2C CH_3O_2C CO_2CH_3

$\xrightarrow{h\nu}$

Supra – antara addition is thermally permitted, but sterically disfavored; it has nevertheless been observed to take place.

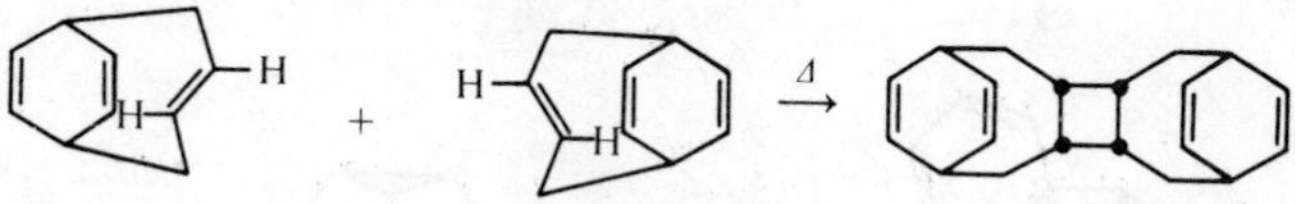

H + H $\xrightarrow{\Delta}$

In the case of the addition of a single bond onto a double bond, the photochemical reaction is allowed if there is inversion on one, but not on the other end of the bond.

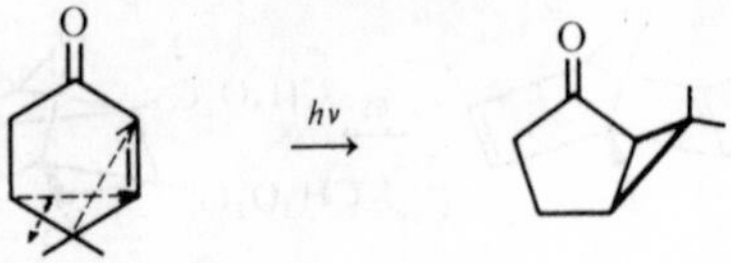

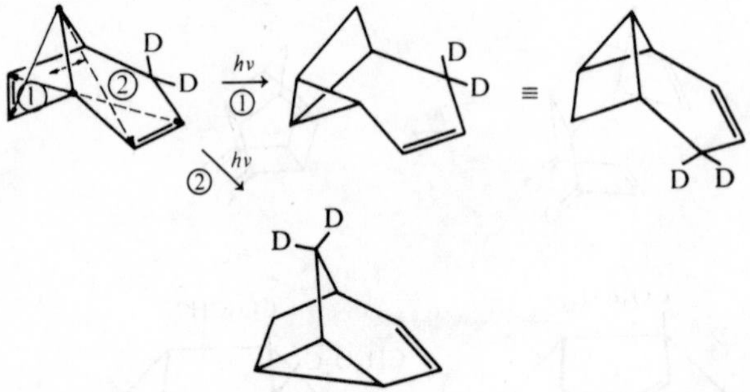

(2 + 4) *addition*

The Diels-Alder reaction can take place thermally by a supra–supra or an antara–antara process. The former is by far the more frequent process; the latter has been observed in the internal cycloaddition of cyclooctatetraenes.

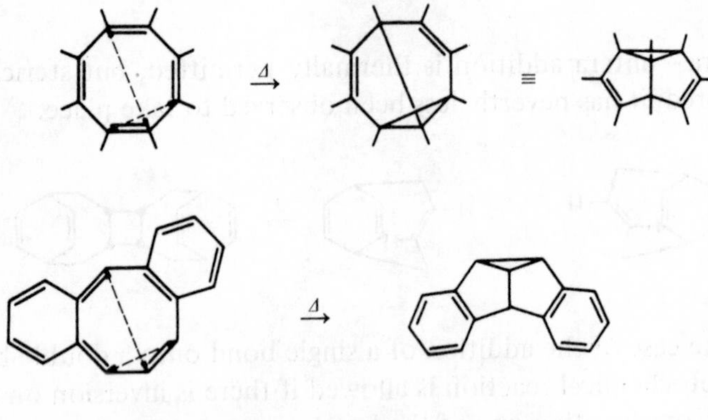

The photochemical rearrangement of hexatriene to bicyclohexene may be considered as a (4 + 2) supra–supra intramolecular addition.

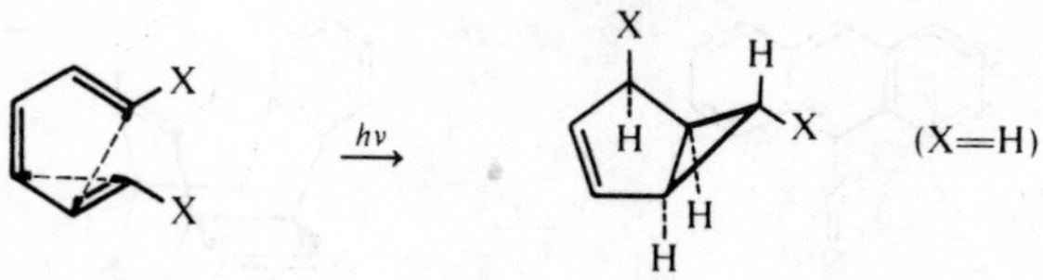

(X=H)

In the case of a substituted hexatriene ($X \neq H$) it may be expected that only one stereoisomer will be formed. This was in fact observed by Dauben et al. for 1, 3, 4, 6-tetraphenylhexatriene:

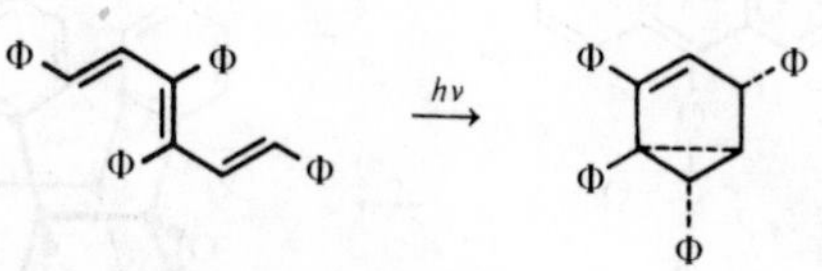

Photochemical rearrangement of 1,2,3,4,5-pentaphenyl cyclohexadiene-1,3 yields only one bicyclohexene isomer. Two mechanisms are conceivable:

a) addition of the 5,6 bond onto the 3,4 double bond;

b) conrotatory ring opening to yield hexatriene, followed by a (4 + 2) supra–antara intramolecular addition.

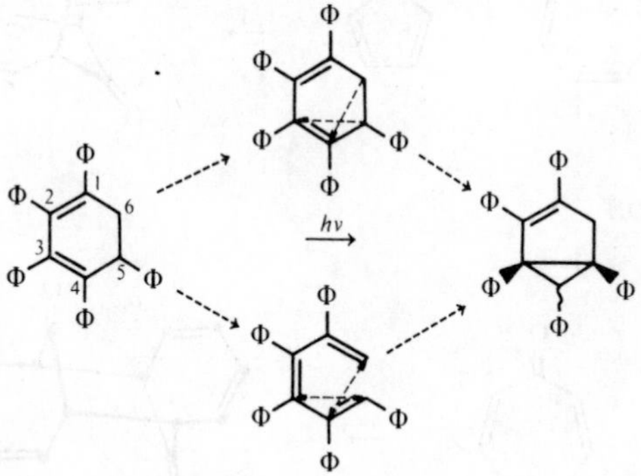

(4 + 4) *addition*

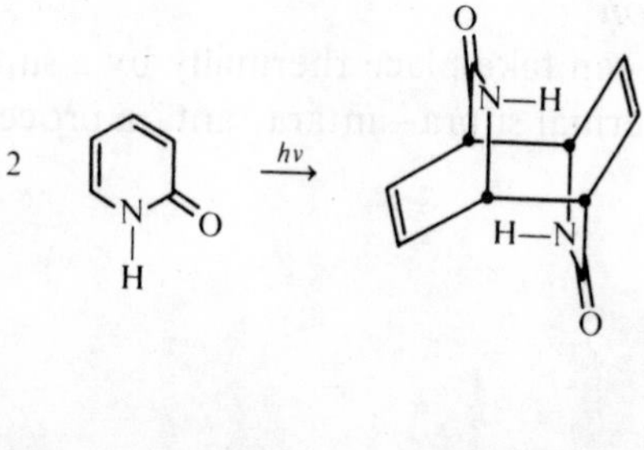

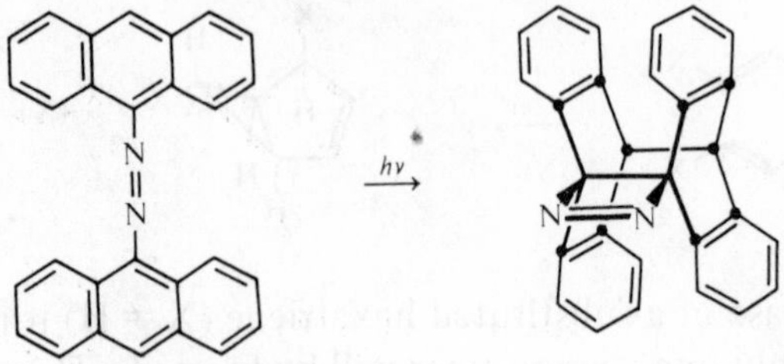

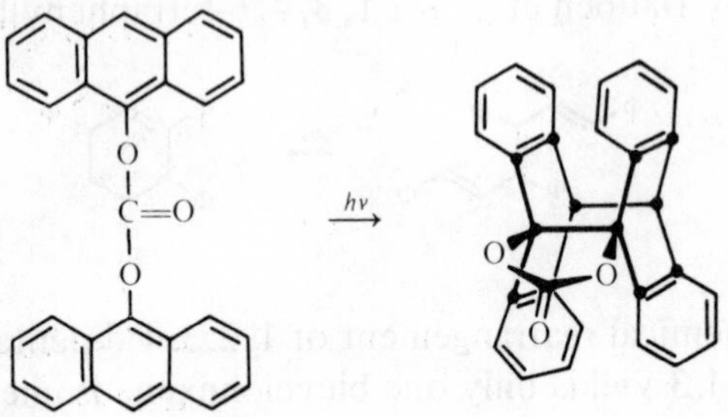

N
N
$h\nu$
N N

O
C=O
O
$h\nu$
O O
O

(6 + 4) *addition*

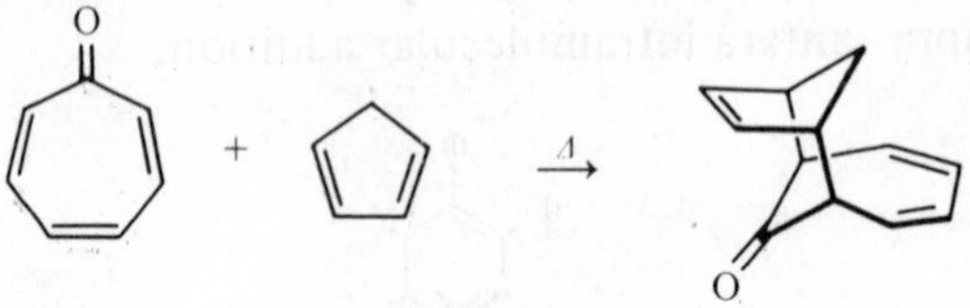

O
+
Δ
O

(6 + 6) *addition*

O
2
$h\nu$
O
O

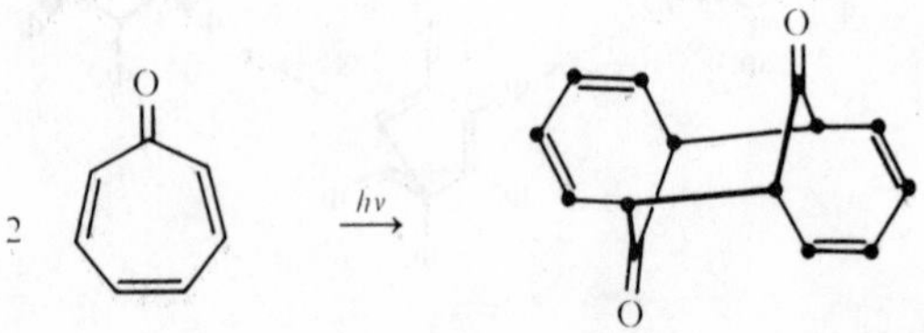

(2 + 2 + 2) *addition*

Such additions can take place thermally by a supra–supra–supra mechanism. A thermal supra–antara–antara process is also allowed:

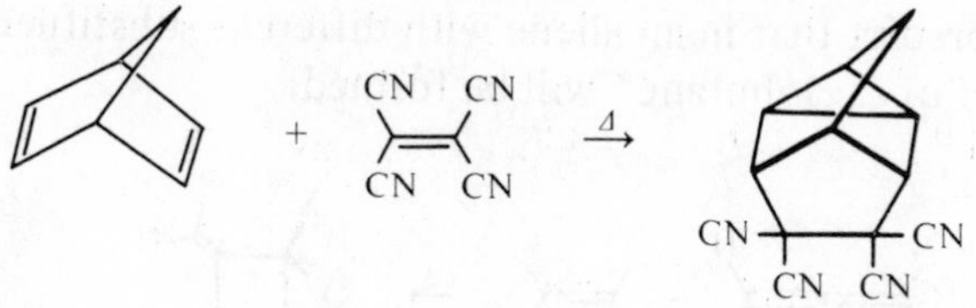

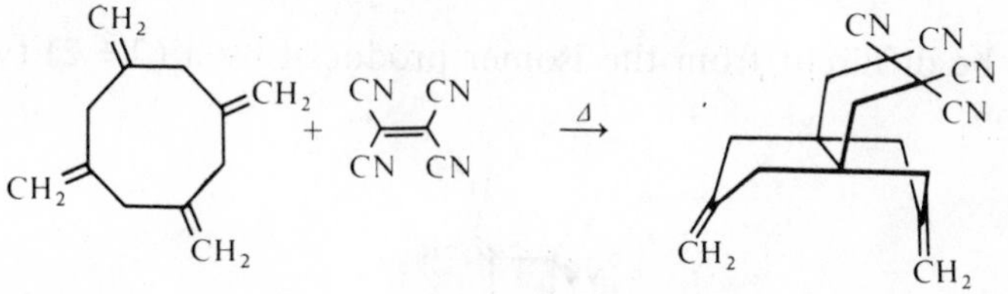

The following thermal reaction, which is a reversed (2 + 2 + 2) addition, has been observed:

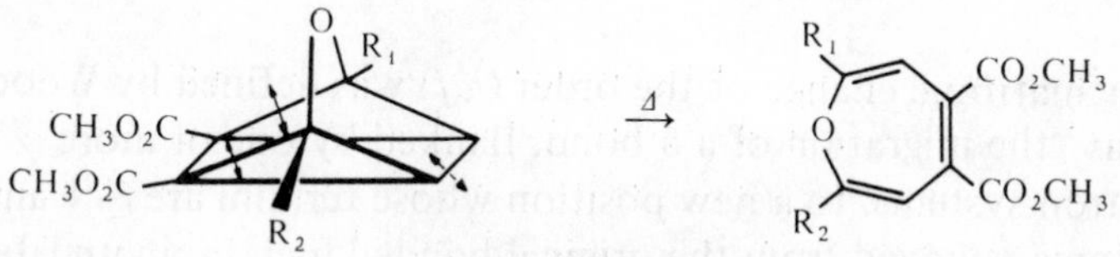

It will be noted that the addition of an allene to a double bond should be thermally forbidden if it is considered as a 2 + 2 reaction; as a matter of fact, this reaction is definitely known to occur. Woodward proposed a (2 + 2 + 2) supra–antara–antara mechanism, in which one π-orbital of each double bond of the allene combines with the π-orbitals of ethylene.

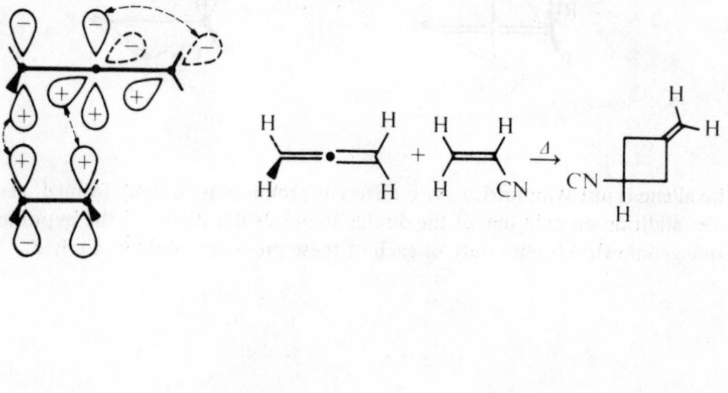

We may predict that in an allene with different substituents, one single isomer of cyclobutane* will be formed:

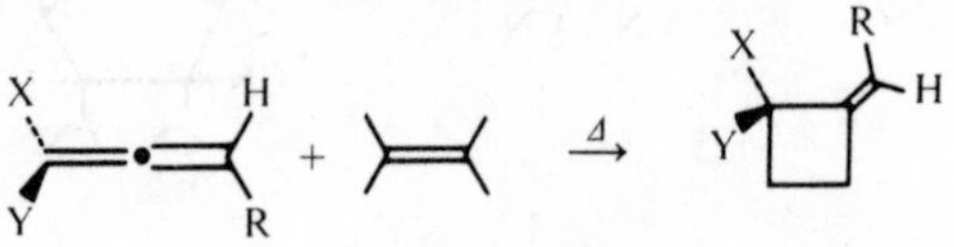

It will be different from the isomer produced by a (2 + 2) type addition:

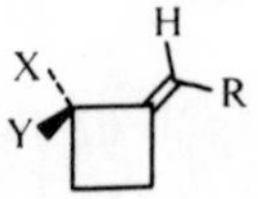

IV. SIGMATROPIC REACTIONS

A sigmatropic change of the order (i, j) was defined by Woodward as "the migration of a σ bond, flanked by one or more π-electron systems, to a new position whose termini are $i-1$ and $j-1$ atoms removed from the original bonded loci, in an uncatalyzed intramolecular process."

Thus, the Cope rearrangement is a sigmatropic change of the order (3,3):

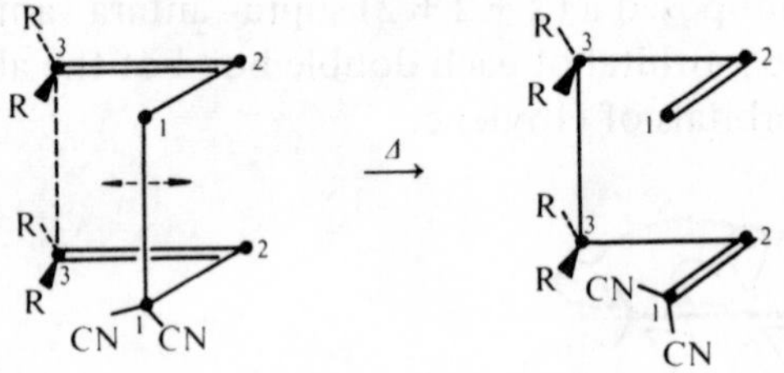

* If the allene is not symmetrical, two different products are usually formed, resulting from the addition on each one of the double bonds of the allene. If the hypothesis were true, one geometrical isomer only of each of these products would be formed.

A bond extremity may be said to migrate according to a suprafacial process if it remains associated with the same face of the π-system in the course of the migration. If the face is changed, we speak of an antarafacial process.

In the most general case of a sigmatropic change of the order (i, j) the migration of a σ-bond may proceed in four different ways:

a) supra–supra: the two extremities of the σ-bond migrate suprafacially;

b) antara–antara: the two extremities of the σ-bond migrate antarafacially;

c) antara–supra and supra–antara (the two may be different): one extremity migrates suprafacially, the other antarafacially.

d) In the particular case of $i = 1$, only two processes – suprafacial and antarafacial – can take place. If we consider the following sigmatropic (1,5) migration:

$$\underset{D}{\overset{C}{>}}\underset{5}{C}{=}\underset{4}{CH}{-}\underset{3}{CH}{=}\underset{2}{CH}{-}\underset{1}{CH}\underset{B}{\overset{A}{<}} \;\longrightarrow\; \underset{D}{\overset{C}{>}}\underset{5}{CH}{-}\underset{4}{CH}{=}\underset{3}{CH}{-}\underset{2}{CH}{=}\underset{1}{C}\underset{B}{\overset{A}{<}}$$

these processes may be represented as follows.

Suprafacial process: the transition state has a plane of symmetry (Figure 6.17).

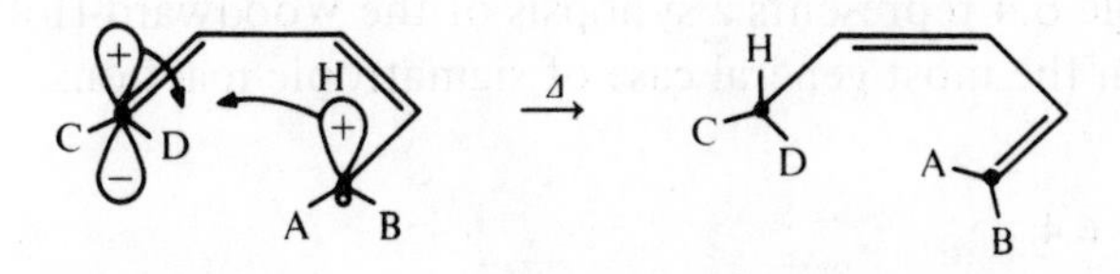

FIGURE 6.17

Antarafacial process: the transition state has a binary axis of symmetry (Figure 6.18).

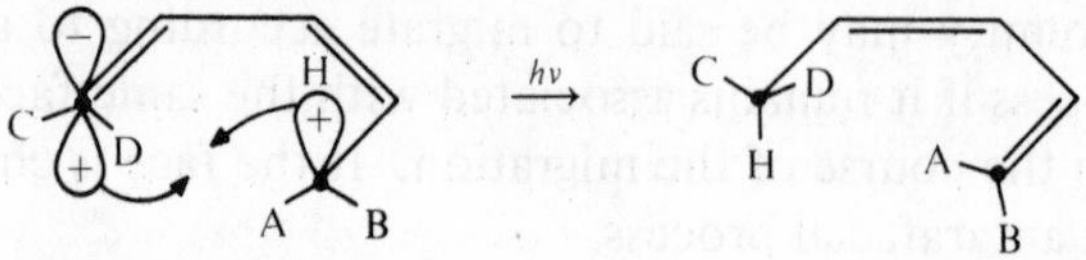

FIGURE 6.18

The transition state may be considered as a hydrogen atom associated with a conjugated polyenic radical; there are p doubly occupied molecular orbitals; the $(p + 1)$-th, which is the highest occupied orbital in the ground state, has a single electron and is present as nodes on all even-numbered carbon atoms. The symmetry of this state is shown in Figure 6.19.

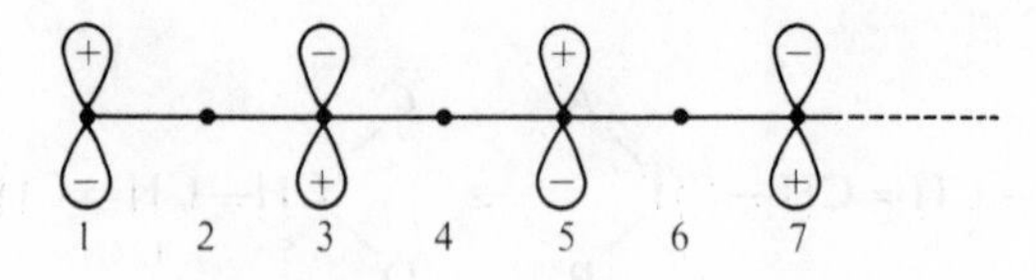

FIGURE 6.19

It is seen, accordingly, that in the case of the above sigmatropic (1,5) reaction the suprafacial process is allowed in the ground state, and is forbidden in the first excited state.

Table 6.4 represents a synopsis of the Woodward-Hoffmann rules in the most general case of sigmatropic reactions.

TABLE 6.4

$i + j$	Thermal	Photochemical
$4q$	antara–supra supra–antara	supra–supra antara–antara
$4q + 2$	supra–supra antara–antara	antara–supra supra–antara

If the reaction is of the order $(1, j)$, these rules can be simplified as shown in Table 6.5.

TABLE 6.5

$1 + j$	j	Thermal	Photochemical
$4q$	3 7	Antarafacial	Suprafacial
$4q + 2$	5	Suprafacial	Antarafacial

1. Notes

An antarafacial process is almost impossible in the case of a small-sized or a medium-sized ring.

An antarafacial process results in a distortion of the polyenic chain; theoretical predictions will only be valid if such a distortion is not too great.

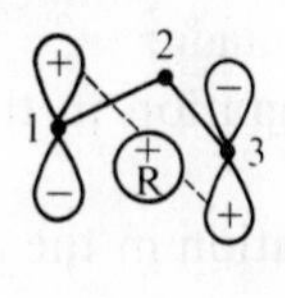

1–3 antarafacial
(retention of configuration)

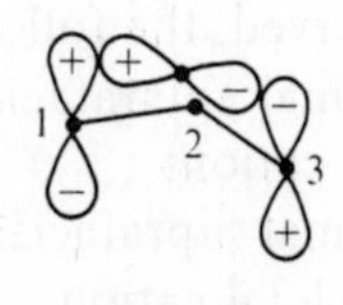

1–3 suprafacial
(inversion of configuration)

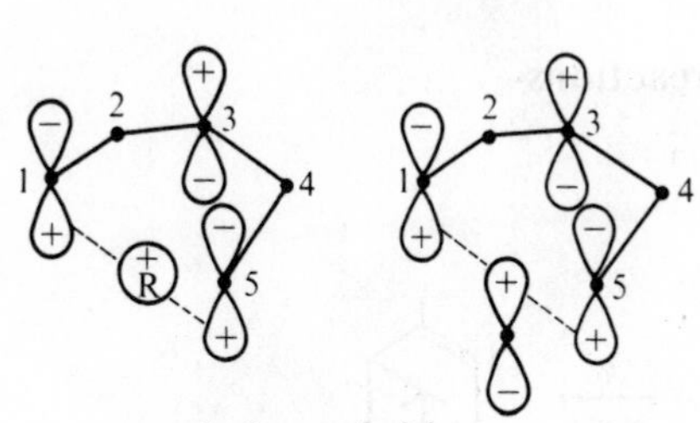

1–5 suprafacial
(retention of configuration)

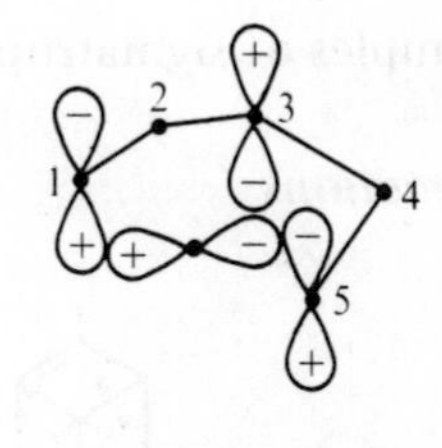

1–5 antarafacial
(inversion of configuration)

FIGURE 6.20

The rules governing sigmatropic reactions are based on the following two assumptions: 1) the bond orbital of the migrating group interacts in its intermediate state with a π-system; 2) in the process of migration, the configuration of the migrating group is retained. If the bond orbital is symmetric (e.g., the *s*-orbital of hydrogen), the configuration must be retained. If the bond orbital is antisymmetric (e.g., the *p*-orbital of a carbon atom), there are two possibilities: if the lobe involved in the original σ-bond is utilized in the formation of the new σ-bond, the configuration is retained; if not, the configuration of the migrating group is inverted (Figure 6.20).

The rules are applicable to ionic systems. Thus, a (1–2) suprafacial slip in a carbonium ion in the ground state is allowed and is very common:

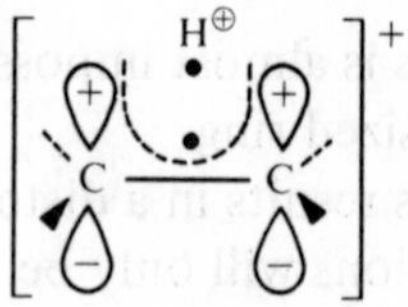

In the case of other cations, for which the migration has not yet been observed, the following predictions can be made:

a) thermal antarafacial process for (1–4) migration in the butenyl-2 cation;

b) thermal suprafacial process for (1–6) migration in the hexadienyl-2,4 cation.

A cyclopropane ring may replace a double bond in the course of a sigmatropic rearrangement.

2. Examples of sigmatropic reactions

(1,3) *reactions*

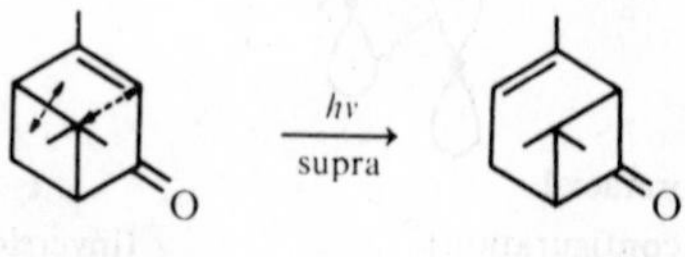

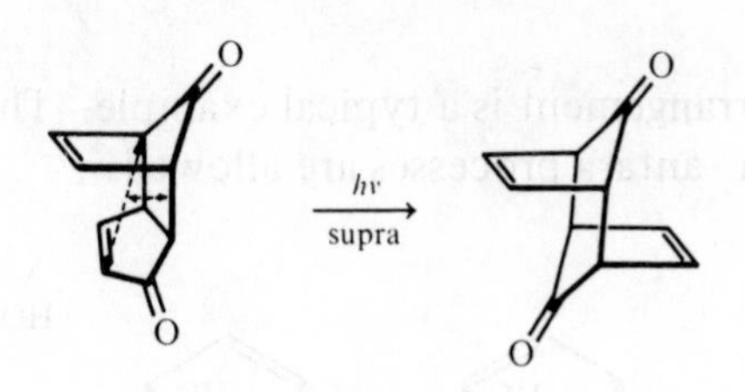

(1,5) *reactions*

Δ
supra

Δ
supra

$$CH_2{=}CH{-}CH{=}CH{-}CH{=}CH_2 \xrightarrow[\text{antarafacial}]{h\nu} CH_2{=}C{=}CH{-}CH{=}CH{-}CH_3$$

(1,7) *reactions*

hν
supra
H

H
hν
supra
H

H D
hν
supra
D

OH
antara
Δ
HO

(3,3) *reactions*

The Cope rearrangement is a typical example. Thermal supra–supra and antara–antara processes are allowed.

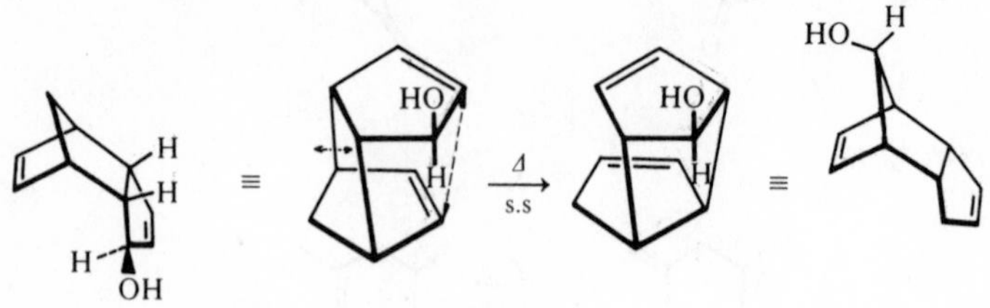

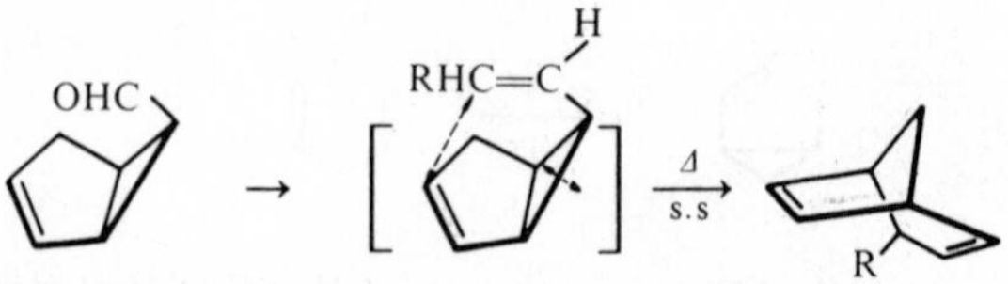

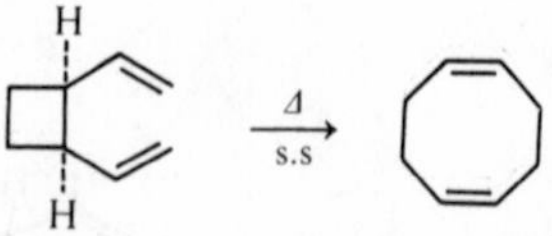

The antara–antara process, though difficult to visualize, has also been observed. The bicyclic ketone which is the product of internal photochemical cyclization of tropolone undergoes a Cope rearrangement by an antara–antara process.

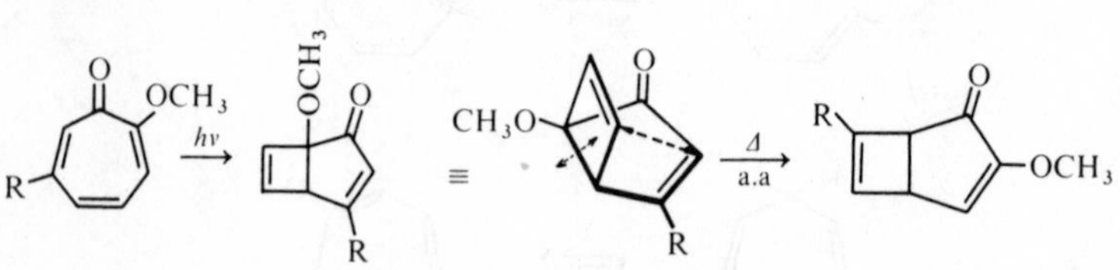

(3,5) *reactions*

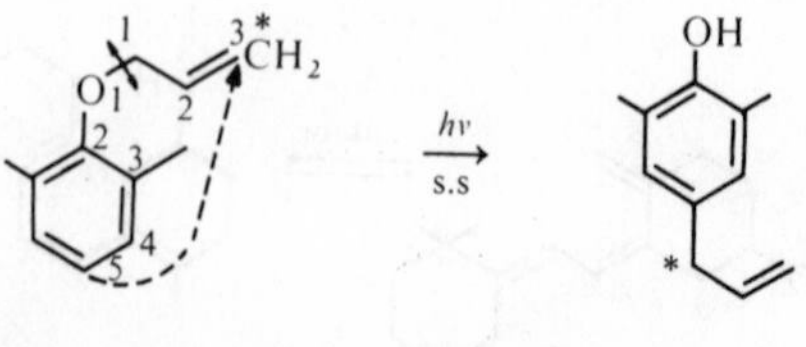

(5,5) *reactions*

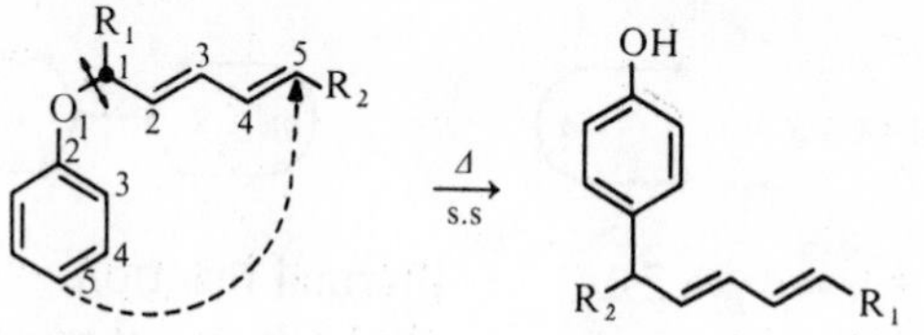

$\xrightarrow[\text{s.s}]{\Delta}$

V. APPLICATIONS OF THE WOODWARD-HOFFMANN RULES TO OTHER TYPES OF REACTION

1. Group transfer

In the schematic reaction represented below, the thermal double group transfer will be theoretically allowed for $m + n = 4q + 2$, while the photochemical transfer will be allowed if $m + n = 4q$, where m and n are the respective numbers of π-electrons.

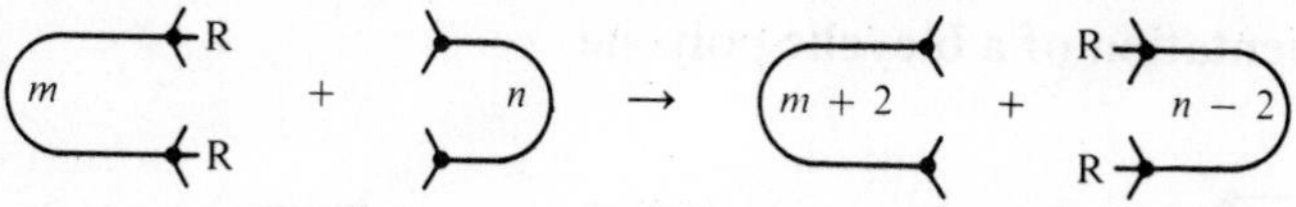

m + n → $m + 2$ + $n - 2$

The rule is valid if $n = 0$ (elimination reaction)

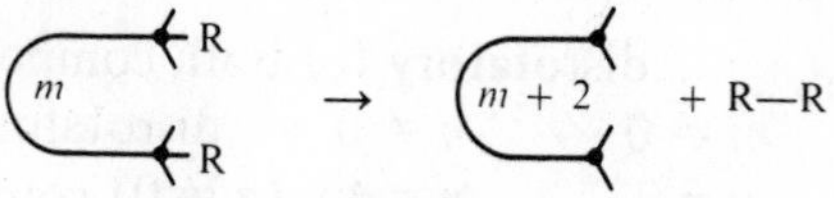

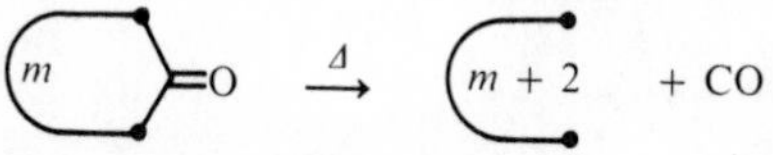

m → $m + 2$ + R—R

m =O $\xrightarrow{\Delta}$ $m + 2$ + CO

$m = 4q + 2$ disrotatory process
$m = 4q$ conrotatory process $q \neq 0$

2. Addition of acetylene to two π-systems

$m + n = 4q + 2$ thermal reaction
$m + n = 4q$ photochemical reaction

3. Decarboxylation of a cyclic ketal

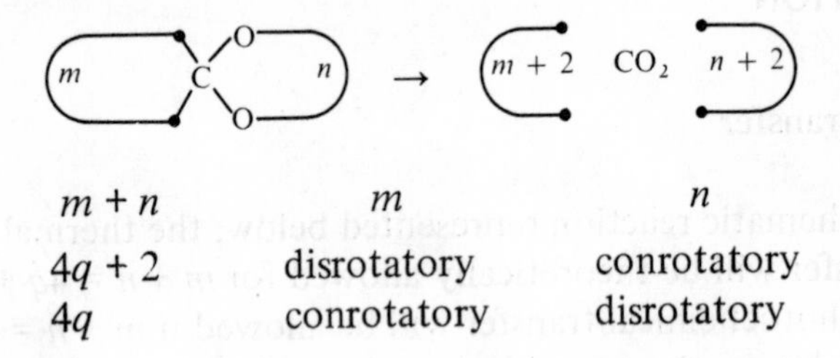

$m + n$	m	n
$4q + 2$	disrotatory	conrotatory
$4q$	conrotatory	disrotatory

4. Fragmentation of a bicyclic polyene

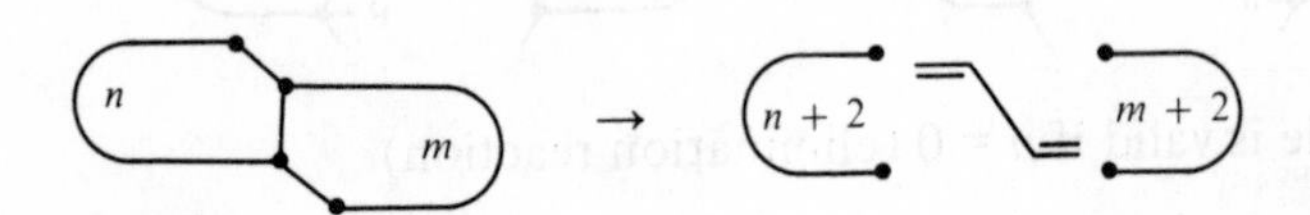

$m + n = 4q + 2$ disrotatory for both components,
$m + n = 4q$ $m \neq 0$ $n \neq 0$ disrotatory or conrotatory,
$m = 0$ $n = 4q$ $(q \neq 0)$ conrotatory for n.

5. Condensation

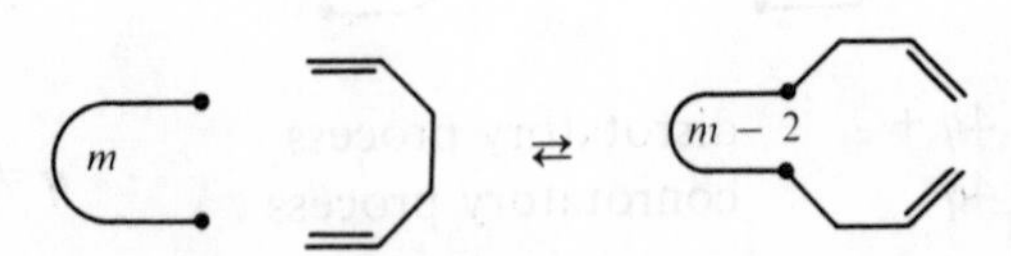

If the system has a plane of symmetry bisecting m

$m = 4q$ thermally allowed
$m = 4q + 2$ photochemically allowed

If the system has a binary symmetry axis, the rules are reversed.

6. Cleavage of oxygen

$\rangle N \to O$ m ⇄ $\rangle N$ O $m - 2$

The thermal reaction is conrotatory for $m = 4q$, and disrotatory for $m = 4q + 2$.

VI. GENERAL COMMENTS ON THE WOODWARD-HOFFMANN RULES

1. The scope of application of the Woodward-Hoffmann rules is not restricted to the types of reactions just discussed. It appears that the rules are valid for all types of concerted reactions.

2. It is very important to remember that correlation diagrams can only be constructed for molecules or systems with suitable symmetries. A symmetry element with respect to which all levels have the same symmetry cannot be employed, neither can a symmetry element which does not bisect the bonds which are closed or opened. In certain cases a mere pseudosymmetry will be sufficient. Thus, in the cyclization of 2-methylbutadiene there is no useful symmetry, but since the electronic symmetry is not considerably affected by the methyl substituent, the considerations valid for unsubstituted butadiene may be applied.

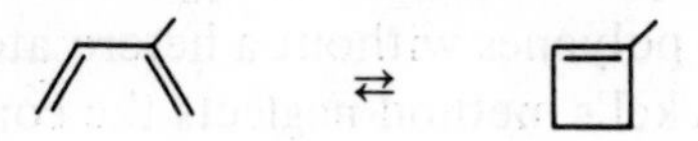

3. If there is no symmetry or pseudosymmetry, it is still possible to predict whether a cycloaddition will be favored or disfavored. The molecular orbitals can be calculated by the Hueckel method. As in the case of a Diels-Alder reaction between two butadienes (p. 117), the criterion is the interaction between the highest occupied orbital of one molecule with the lowest vacant orbital of the other: if the interaction is bonding, the reaction is "allowed," if it is antibonding, the reaction is "forbidden."

4. We have seen that the rules do not distinguish between two conrotatory or two disrotatory processes. If a given system can give two reactions, e.g., a cycloaddition or a cyclization, the favored reaction cannot be identified by the Woodward-Hoffmann rules, but only by other, mainly steric, considerations.

5. The Woodward-Hoffmann rules make it possible to predict whether or not a given reaction will be favored. Under very high energy conditions, a disfavored reaction may take place. It is estimated that the additional energy which is required to violate the rules is 15 kcal·mole^{-1}.

6. Since the rules apply to concerted reactions only, the experimentally observed stereochemistry of the reaction may be in contradiction with the rules valid for a multistage process. In photochemical reactions the rules are valid only for the first excited singlet state.

7. The rules are based on the highest occupied molecular orbital. This is justified, since if the calculation is carried out by taking into account all the orbitals, it is found that the term originating from the highest occupied molecular orbital is by far the largest.

8. In electrocyclic reactions, only the symmetry of the open-chain polyene is considered. This procedure is justified by the principle of microreversibility of concerted reactions.

9. Since the rules are based on the method of Hueckel, their validity does not go beyond that of this approximation; in particular, they apply to polyenes without a hetero atom in the molecule. Since Hueckel's method neglects the contribution of the *n*-electrons of the hetero atom, it is not always reliable when applied to this type of compound.

10. Other interpretations have been offered /12–17/ to account for the stereospecificity of concerted reactions.

Bibliography

1. Hoffmann, R. and R. B. Woodward. – Accounts of Chemical Research, **1** : 17. 1968.
2. Berson, J. A. – Accounts of Chemical Research, **1** : 152. 1968.
3. Millie, P. – Bull. Soc. Chim. France, p.4031. 1966.
4. Steinmetz, R. – Fortsch. Chem. Forsch. (Photochimie), **7** : 445. 1967.
5. Kwart, H. and K. King. – Chem. Rev., **68** : 415. 1968.
6. Vollmer, J. J. and K. L. Servis. – J. Chem. Educ., **45** : 214. 1968.
7. Huisgen, R. – Angew. Chem. (Intern. Ed.), **7** : 321. 1968.
8. Woodward, R. B. Aromaticity, Special Publication No.21, p.217. The Chemical Society, London. 1967.
9. Gill, G. B. – Quart. Rev., **22** : 338. 1968.
10. Miller, S. I. Advances in Physical Organic Chemistry. Edited by V. Gold. **6** : 185. 1968.
11. Frey, H. M. Ibid., **4** : 147. 1966.
12. Fukui, K. – Tetrahedron Letters, p.2009. 1965. Bull.Chem. Soc. Japan, **39** : 498. 1966.
13. Fukui, K. and H. Fujimoto. – Ibid., **39** : 2116. 1966.
14. Longuet-Higgins, H. C. and E. W. Abrahamson. – J. Am. Chem. Soc., **87** : 2045. 1965.
15. Zimmerman, H. E. – Ibid., **88** : 1564, 1566. 1966.
16. Dewar, M. J. S. – Tetrahedron Suppl., **8** : 75. 1967.
17. Salem, L. – J. Am. Chem. Soc., **90** : 543, 553. 1968.

Chapter 7

PHOTOCHEMISTRY OF ALDEHYDES AND KETONES

I. INTRODUCTION

Photochemical reactions of carbonyl compounds account for a very large proportion of all known photochemical reactions. This is due both to the fact that the carbonyl group can be selectively excited, and to the fact that it is highly reactive. Its reactivities are different in the ground and in the excited states.

The $(n \rightarrow \pi^*)$ transition is the main transition in the photochemistry of carbonyl compounds. The wavelength of this transition varies with the particular class of carbonyl compounds, as follows:

a) saturated aldehydes and ketones: λ_{max} 2800–3000 Å;
b) α, β-unsaturated aldehydes and ketones: λ_{max} 3000–3500 Å;
c) aromatic aldehydes and ketones: λ_{max} 3000–3500 Å.

The intensity (ϵ) of the $(n \rightarrow \pi^*)$ band is invariably very low: ~20 for aliphatic and ~100 for aromatic aldehydes and ketones.

II. PRIMARY PHOTOCHEMICAL REACTIONS OF ALDEHYDES AND KETONES

1. Intramolecular processes

The three main types of intramolecular reactions given by carbonyl compounds are decomposition, reduction, and rearrangement.

A) *Decomposition reactions*

According to Norrish, carbonyl compounds give three types of decomposition reactions.

Type I: Decomposition into free radicals.

A carbonyl compound $R_1{-}\underset{\underset{O}{\|}}{C}O{-}R_2$ can decompose into free radicals in three ways:

$$R_1{-}\underset{\underset{O}{\|}}{C}{-}R_2 \xrightarrow{h\nu} \begin{cases} R_2\cdot + \cdot\underset{\underset{O}{\|}}{C}R_1 & (Ia) \\ R_1\cdot + \cdot\underset{\underset{O}{\|}}{C}{-}R_2 & (Ib) \\ R_1\cdot + R_2\cdot + CO & (Ic) \end{cases}$$

If $R_1 = R_2$, reactions (Ia) and (Ib) are identical.

If $R_1 = H$ (aldehydes), the compound will decompose mainly according to (Ia).

If $R_1 \neq R_2$, one of the two processes (Ia) and (Ib) will be favored over the other; the favored process will involve the rupture of the weaker bond and formation of more stable radicals. The lower the energy of the photon, the more conspicuous the selectivity.

It is important to remember that the free radical decomposition taking place in solvents at room temperature, or even at elevated temperatures, is often prevented by the recombination of the radicals thus formed ("cage effect" of the solvent). In the vapor phase, the decomposition reactions become conspicuous and the free radicals formed participate in secondary reactions.

Type I decompositions are typical of ketones in which R_1 or R_2 have no γ-hydrogen; if γ-hydrogen is present, type II reactions become possible.

Type II:

Reactions of this type consist in the decomposition into molecules of aliphatic aldehydes and ketones with a γ-hydrogen in the alkyl chain. The mechanism of this reaction involves the cleavage of the γ-hydrogen by the excited carbonyl (six-membered cyclic intermediate):

$$R_1\text{—}\overset{\overset{\displaystyle O}{\|}}{C}\text{—}CH_2CH_2\text{—}CH\begin{matrix}R_2\\ R_3\end{matrix} \xrightarrow{h\nu} \text{[cyclic intermediate: } R_1\text{—C(=O}\cdots\text{H}\cdots\text{C}R_2R_3\text{)]}$$

$$\longrightarrow R_1\text{—}\overset{\overset{\displaystyle OH}{|}}{C}=CH_2 + CH_2=C\begin{matrix}R_2\\ R_3\end{matrix}$$

$$\downarrow$$

$$R_1\text{—}\overset{\overset{\displaystyle O}{\|}}{C}\text{—}CH_3$$

The ketone is obtained in its enol form, as indicated by the results of IR spectroscopy.

Type II reactions take place just as readily in the liquid as in the vapor phase. Unsaturated aliphatic aldehydes and ketones, halogen-substituted aldehydes and ketones, and 1,2-diketones do not seem to give this reaction. If R_1 is an aromatic grouping, on the other hand, the reaction may take place. The reaction is given by cyclic ketones and by other carbonyl compounds such as esters, acid anhydrides and acid amides (Chapter 4). It now seems certain that the reaction mechanism involves the triplet state $^3(n \rightarrow \pi^*)$ of the ketone.

It has been shown by Pitts et al. that the γ-proton may form part of a cyclopropyl group:

$$CH_3COCH_2CH_2\text{-}(C_3H_5) \xrightarrow{h\nu} CH_3C(OH)=CH_2 + (C_3H_4)=CH_2$$

Type III:

Similarly to type II reactions, this reaction is a photocycloelimination which yields one aldehyde molecule and one olefin molecule. The mechanism probably involves cleavage of β-hydrogen, but has been less thoroughly studied than the mechanism of type II reactions.

$$R{-}\overset{O}{\overset{\|}{C}}{-}CH(CH_3)_2 \xrightarrow{h\nu} R{-}CHO + CH_2{=}CH{-}CH_3$$

Unlike type II reactions, type III reactions are sensitive to the wavelength of the irradiating light.

B) *Reduction reactions*

During intramolecular reduction of a carbonyl group by the withdrawal of a hydrogen atom in a γ-position to the carbonyl, the cyclic intermediate compound is the same as in type II decomposition; in such a case, an olefin molecule is not eliminated, but the intermediate biradical compound becomes recyclized.

Depending on the identities of R_2 and R_3, different kinds of cyclization following the rearrangement of the radical are possible.

This mechanism, which involves allylic rearrangement of the intermediate radical, is not certain, since it fails to account for the retention of the configuration which is observed in the photochemical reduction of 2,6-dimethyloctene-7-one-3; two concerted processes, a six-member and a four-member process, with stereospecific cleavage of the allylic hydrogen atom, have been postulated for this case.

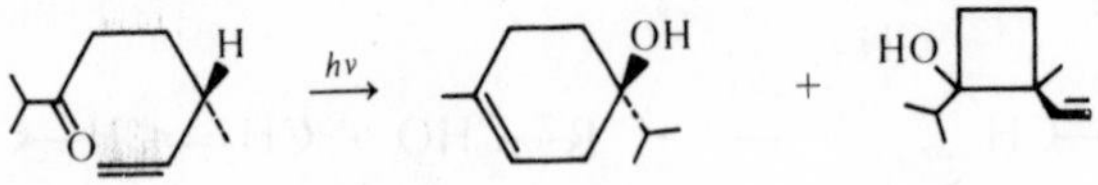

Other types of intramolecular photochemical reduction, which are not caused by the cleavage of the γ-hydrogen atom, have been observed:

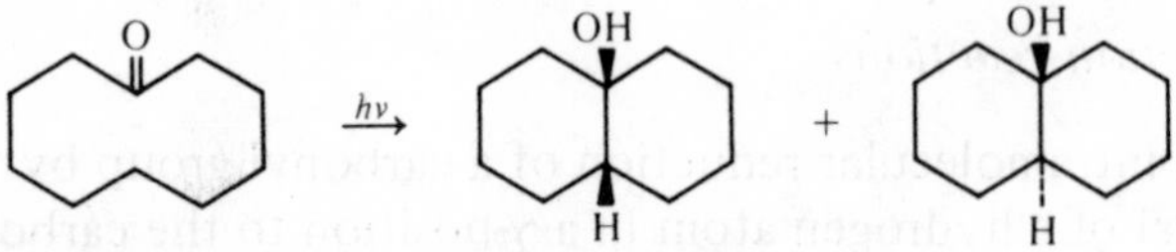

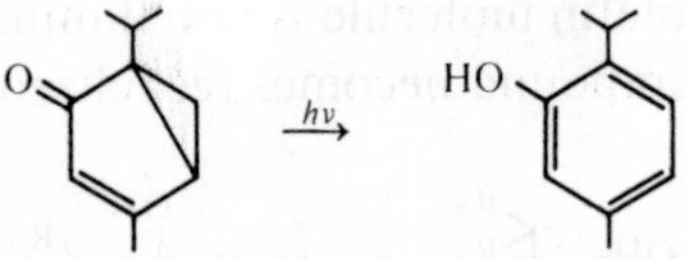

C) *Rearrangement reactions*

Intramolecular rearrangements without reduction of the carbonyl group include *cis-trans* isomerization of unsaturated compounds, ring closure, ring opening, and ring contraction.

The *cis-trans* isomerization of α, β-unsaturated aldehydes and ketones is usually the main photochemical reaction given by these compounds. The double bond may or may not migrate to the β,γ-position.

Ring closure may be observed in the case of unsaturated aliphatic ketones, with a quaternary carbon atom in the γ-position.

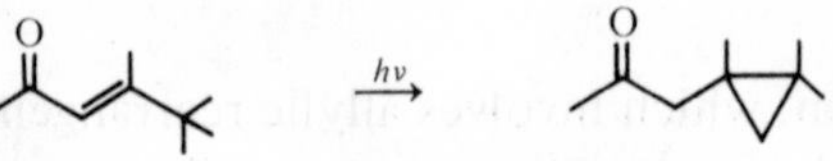

Ring closure reactions have also been observed for a number of unsaturated ketones; such reactions yield bicyclic or polycyclic ketones.

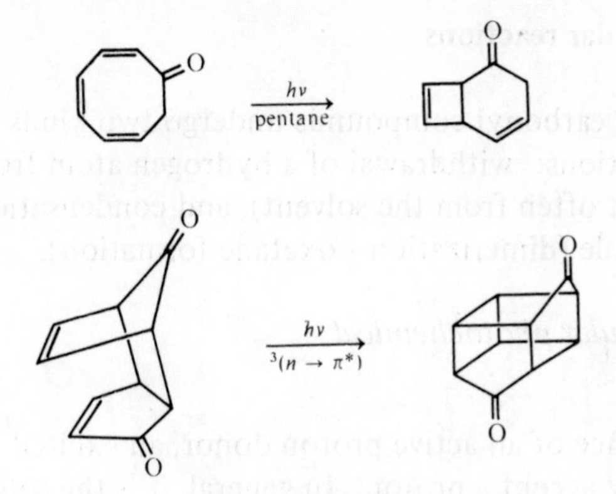

Ring opening is a very well known photochemical reaction. It includes, in particular, the rearrangement of cyclopropyl ketones to ethylenic ketones and, in general, of ring ketones to ethylenic open-chain ketones; the epoxide ring of α, β-epoxy ketones may also open, with the formation of β-diketones.

These reactions also include ring contraction of 5, 6, 7 or 8-membered ring ketones.

2. Intermolecular reactions

In principle, carbonyl compounds undergo two kinds of intermolecular reactions: withdrawal of a hydrogen atom from another molecule (most often from the solvent), and condensation with another molecule (dimerization – oxetane formation).

A) *Intermolecular photochemical reductions*

In the presence of an active proton donor, an excited carbonyl compound may accept a proton. In general, it is the solvent which acts as a proton donor and intermolecular photochemical reductions are observed in liquid phase only. Again, it would seem that the reactive species is the triplet state $^3(n \rightarrow \pi^*)$ of the ketone; if the state $^3(\pi \rightarrow \pi^*)$ is the lowest triplet state, there is no photochemical reduction, while an excited singlet state is as a rule responsible for other kinds of reactions.

Aromatic ketones and ring ketones are more readily reduced than aliphatic ketones. If type II decomposition is possible, the photochemical reduction reaction is usually negligible.

In the presence of a proton-donating solvent aromatic ketones are usually converted to pinacols. Thus, benzophenone dissolved in isopropanol gives a quantitative yield of benzopinacol.

$$\Phi_2C{=}O \xrightarrow[(CH_3)_2CH{-}OH]{h\nu} \Phi_2\dot{C}{-}OH \rightarrow \Phi_2C(OH){-}C(OH)\Phi_2$$

B) *Photochemical dimerization of a, β-unsaturated carbonyl compounds* /10/

Condensation of two molecules of an α,β-unsaturated carbonyl compound with a cyclobutane dimer is well known to take place in the liquid phase. The resulting dimer may be a head-to-head or a head-to-tail dimer:

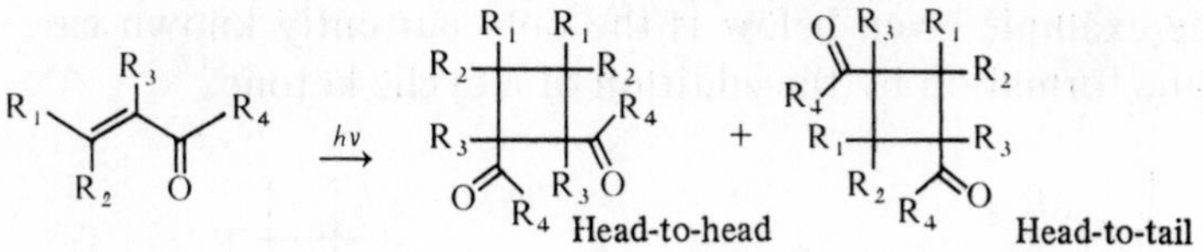

Head-to-head Head-to-tail

These dimerizations seem to originate from ($\pi \rightarrow \pi^*$) excited states, since they have also been observed in the absence of a carbonyl group.

C) *Photochemical addition to olefins* /10,11/

Aliphatic and aromatic aldehydes and ketones in their excited state add onto olefins to give oxetanes (Paterno-Buechi reaction), while cyclic ketones and unsaturated ketones do not react under these conditions. Again it seems that $^3(n \rightarrow \pi^*)$ is the excited reactive state; the reactivity shows considerable variations with the structures of the carbonyl compound and of the olefin.

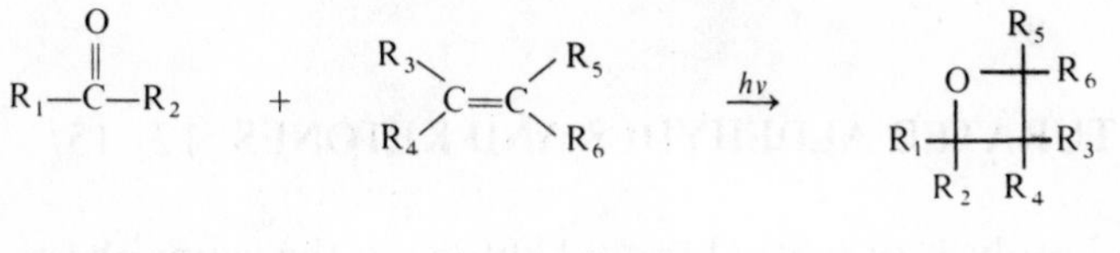

The reaction is given by cyclic olefins:

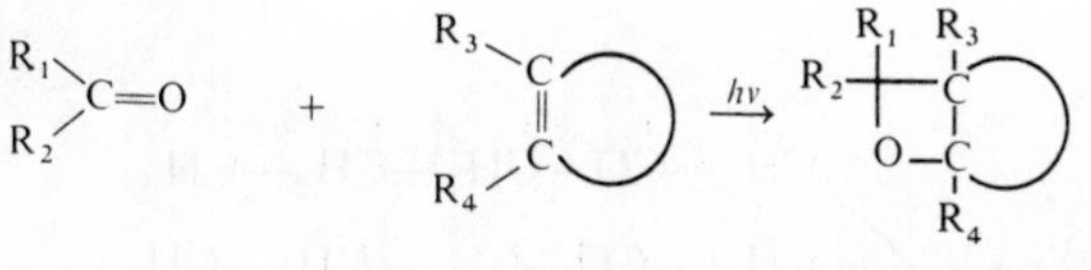

On the other hand, α,β-unsaturated carbonyl compounds yield cyclobutane and not oxetane compounds.

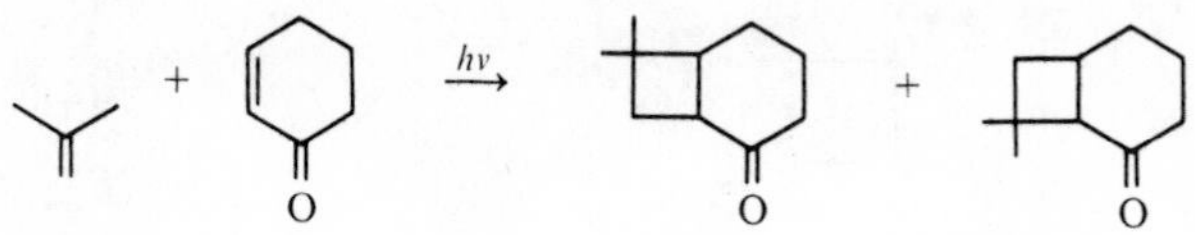

The example given below is the only currently known case of oxetane formation by the addition of a cyclic ketone:

$$\text{(perfluorocyclobutanone)} + CF_3{-}CF{=}CF_2 \xrightarrow{h\nu} \text{(fluorinated oxetane with } CF_3\text{)}$$

Intramolecular oxetane formation is also known, as well as formation of unsaturated oxetane by the addition of a carbonyl to an acetylenic compound; in the latter case the oxetane is unstable and cannot be isolated.

$$\Phi{-}\overset{O}{\overset{\|}{C}}{-}H + R{-}C{\equiv}C{-}R \xrightarrow{h\nu} \left[\text{oxetene } (\Phi, R, R)\right] \rightarrow \Phi(H)C{=}C(R){-}C(O){-}R$$

III. SATURATED ALDEHYDES AND KETONES /12–15/

The photolysis of methyl propyl ketone in the vapor phase, which is a classical example, involves four different primary processes:

$$CH_3COCH_2CH_2CH_3 \xrightarrow{h\nu} \begin{cases} CH_3{-}\dot{C}O + \dot{C}H_2{-}CH_2{-}CH_3 & (1) \\ CH_3\cdot + \dot{C}O{-}CH_2{-}CH_2{-}CH_3 & (2) \\ CH_2{=}CH_2 + CH_3{-}\underset{OH}{\underset{|}{C}}H{=}CH_2 & (3) \\ \text{1-methylcyclobutanol (OH)} & (4) \end{cases}$$

Secondary processes include the decomposition of $CH_3-\dot{C}O$ and $CH_3-CH_2-CH_2-\dot{C}O$ radicals with formation of carbon monoxide and methyl and propyl radicals; recombination or rearrangement of different alkyl radicals, with formation of various hydrocarbons, and the formation of acetone from the enol produced in reaction (3) above.

Saturated aliphatic aldehydes behave in a somewhat different manner, in particular as regards the intramolecular elimination of CO. The vapor phase photolysis of butyraldehyde is an example of this difference. The following primary processes have been studied.

$$H-C(=O)-CH_2-CH_2-CH_3 \xrightarrow{h\nu} \begin{cases} H\dot{C}O + \cdot CH_2-CH_2-CH_3 \\ CO + CH_3-CH_2-CH_3 \\ CH_2{=}CH_2 + CH_2{=}CHOH \rightarrow CH_3-CHO \\ \dot{C}H_3 + \cdot CH_2-CH_2-CHO \\ H-C(=O)-H + CH_3-CH{=}CH_2 \end{cases}$$

In the case of aldehydes and ketones with a cyclopropane ring in the α-position to the carbonyl group, the main primary process seems to be isomerization to an α,β-unsaturated compound by ring opening:

$$\triangle\!-CHO \rightarrow \begin{cases} CH_3-CH{=}CH-CHO \\ CO + \triangle \\ CO + CH_3-CH{=}CH_2 \end{cases}$$

$$\triangle\!-C(=O)-CH_3 \rightarrow \begin{cases} CH_3-CO-CH{=}CH-CH_3 \\ \dot{C}H_3 + \dot{C}_3H_5 + CO \end{cases}$$

On the other hand, ring opening is not observed in methyl cyclobutyl ketone, and the main process in the vapor phase is CO formation:

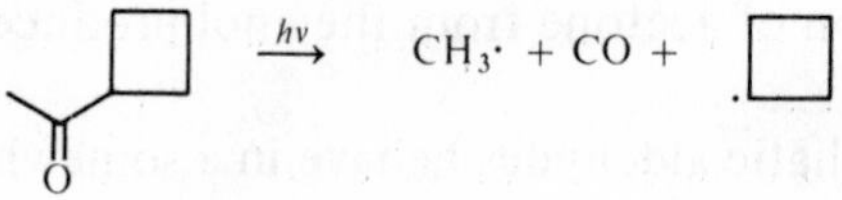

$$\xrightarrow{h\nu} CH_3\cdot + CO +$$

Bicyclic ketones also undergo opening of the cyclopropane ring:

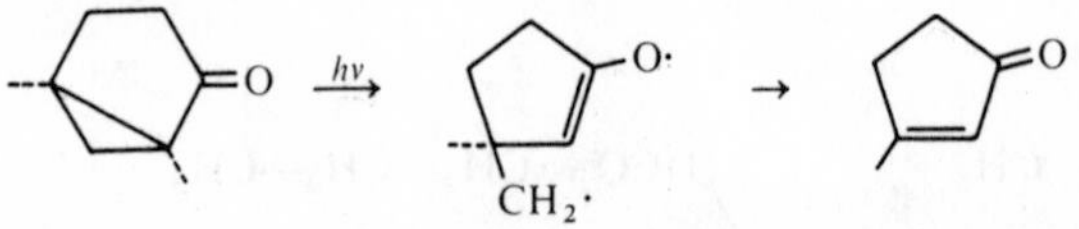

$$\xrightarrow{h\nu} \quad O\cdot \quad CH_2\cdot \quad \rightarrow \quad O$$

According to Pitts et al., the opening of this ring is facilitated by the overlap of the π^* orbital and the neighboring σ orbital of cyclopropane.

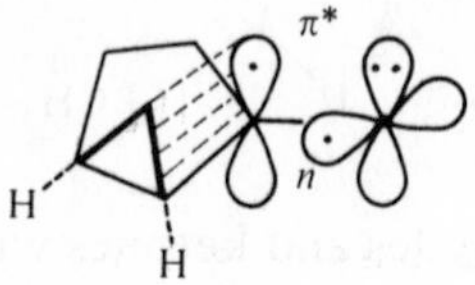

π^* n H H

The presence of a cyclopropyl group conjugated with the carbonyl inhibits type I decomposition; the same effect is produced by a double bond, as will be seen below. If the cyclopropane is in position β or γ to the ketone group, decomposition will be the main process.

Cyclic ketones in the vapor phase usually yield carbon monoxide and a cyclic hydrocarbon with one carbon atom less:

$$(CH_2)_n \; C{=}O \xrightarrow{h\nu} CO + (CH_2)_n$$

Other processes which have been observed are isomerization to an open chain ethylenic aldehyde and decomposition:

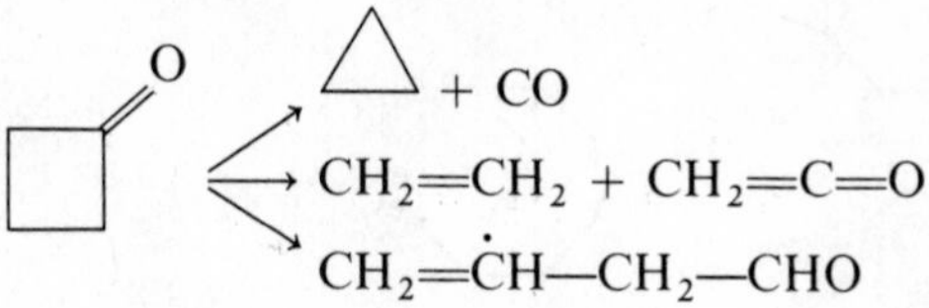

$+ CO$

$CH_2{=}CH_2 + CH_2{=}C{=}O$

$CH_2{=}\dot{C}H{-}CH_2{-}CHO$

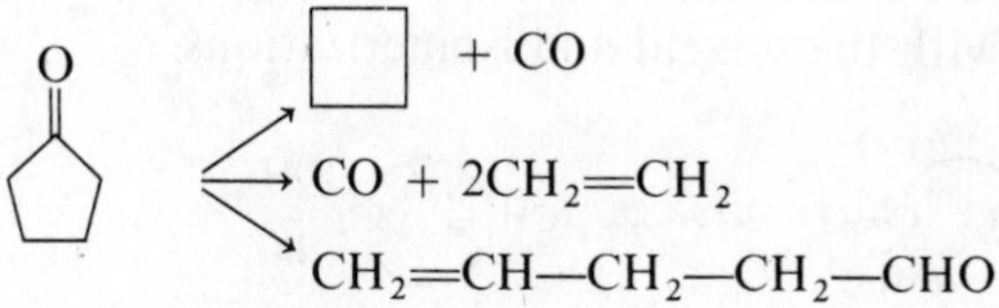

$+ CO$

$CO + 2CH_2{=}CH_2$

$CH_2{=}CH{-}CH_2{-}CH_2{-}CHO$

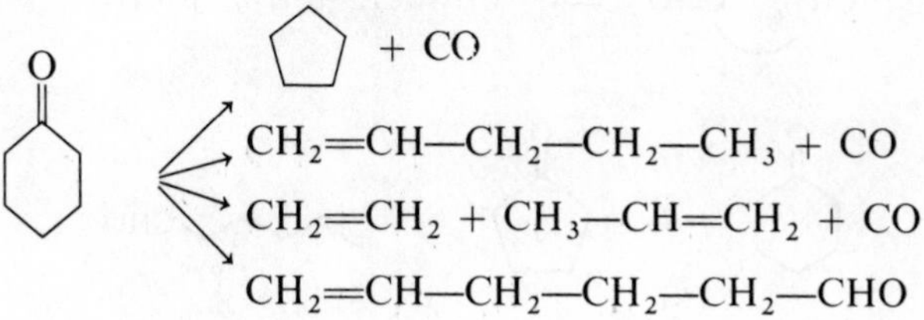

$+ CO$

$CH_2{=}CH{-}CH_2{-}CH_2{-}CH_3 + CO$

$CH_2{=}CH_2 + CH_3{-}CH{=}CH_2 + CO$

$CH_2{=}CH{-}CH_2{-}CH_2{-}CH_2{-}CHO$

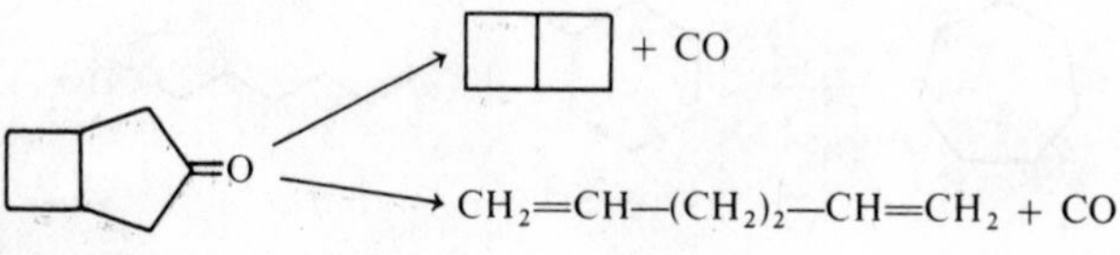

$=O$

$+ CO$

$CH_2{=}CH{-}(CH_2)_2{-}CH{=}CH_2 + CO$

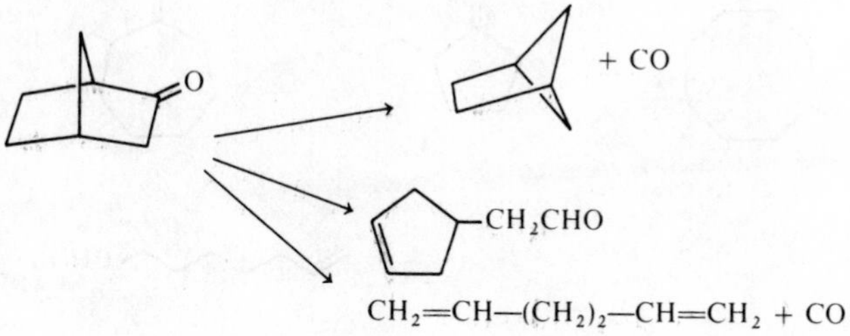

O

$+ CO$

$-CH_2CHO$

$CH_2{=}CH{-}(CH_2)_2{-}CH{=}CH_2 + CO$

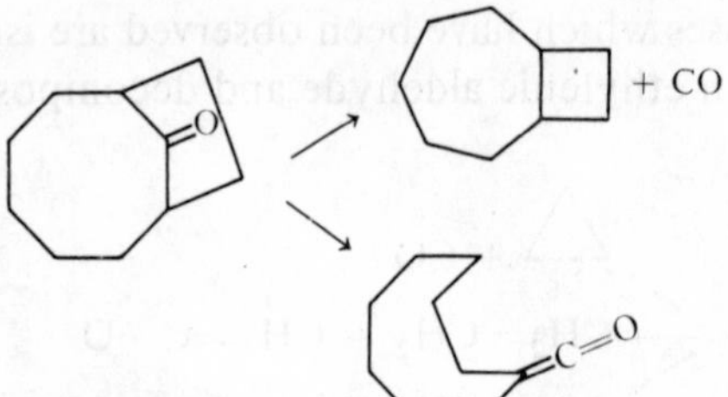

+ CO

C=O

In liquid phase the decomposition of cyclic ketones into a cyclic hydrocarbon and CO is not favored; the main processes are bimolecular reactions with the solvent and isomerizations.

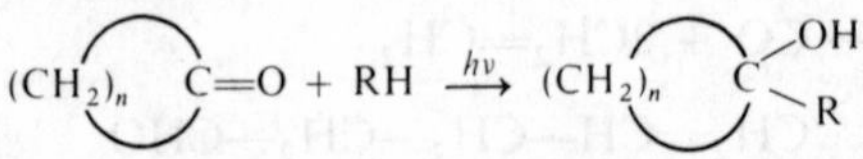

$$(CH_2)_n \; C{=}O + RH \xrightarrow{h\nu} (CH_2)_n \; C(OH)(R)$$

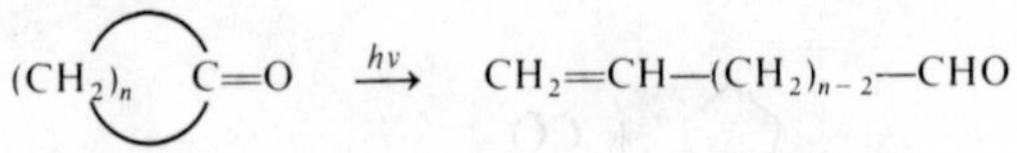

$$(CH_2)_n \; C{=}O \xrightarrow{h\nu} CH_2{=}CH{-}(CH_2)_{n-2}{-}CHO$$

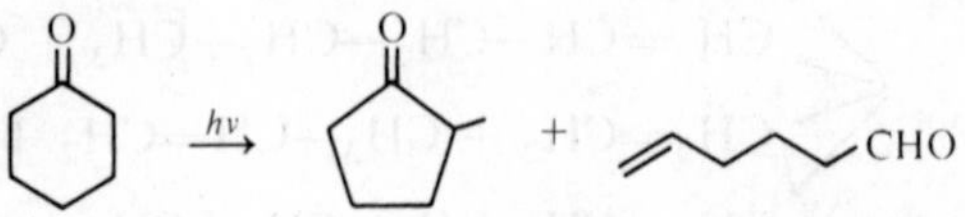

O $\xrightarrow{h\nu}$ O + CHO

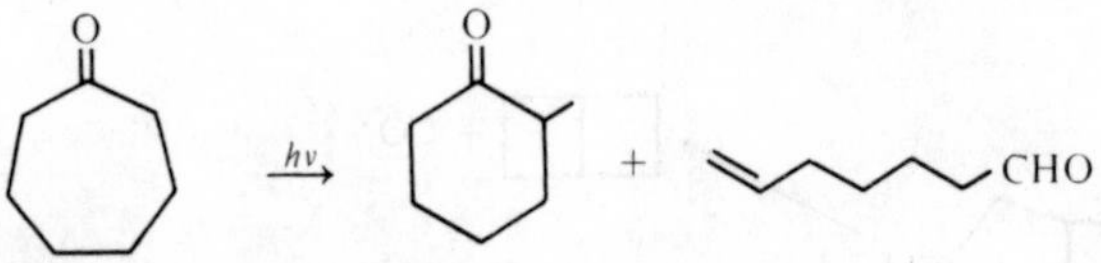

O $\xrightarrow{h\nu}$ O + CHO

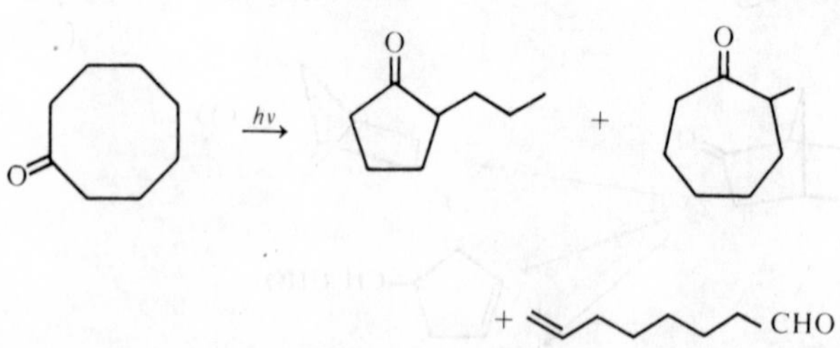

O $\xrightarrow{h\nu}$ O + O

+ CHO

IV. α, β-UNSATURATED ALDEHYDES AND KETONES /16–25/

The presence of an α,β-double bond in position α to the carbonyl strongly suppresses decomposition reactions. The main reaction of open-chain compounds is the migration of the double bond to the β, γ-position (probably by a photoenolization mechanism).

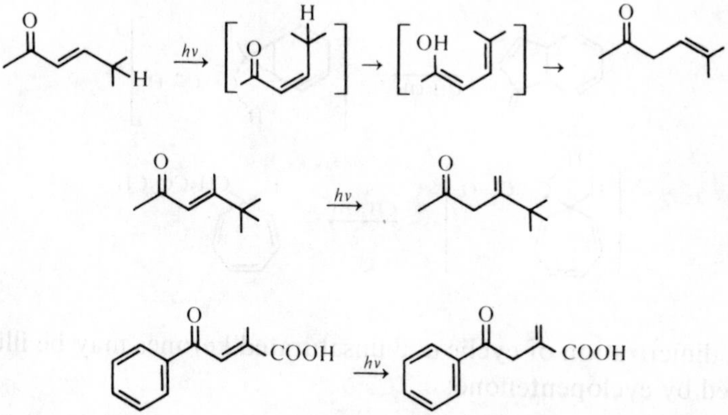

The photoenolization mechanism is not operative in the isomerization of 10-α-testosterone:

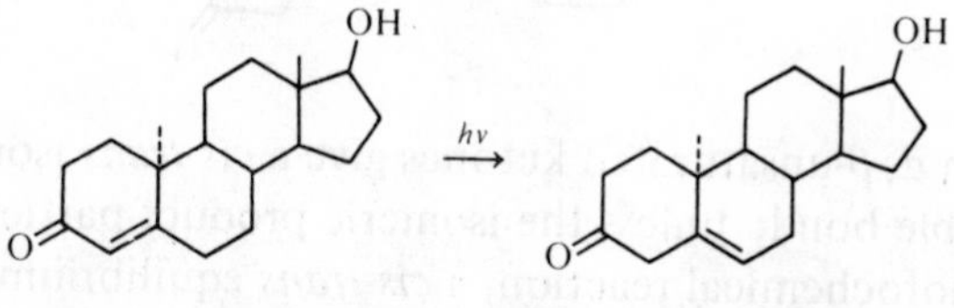

If the molecule includes another double bond, intramolecular cyclization may take place; an example is the photochemical cyclization of citral and of carvone.

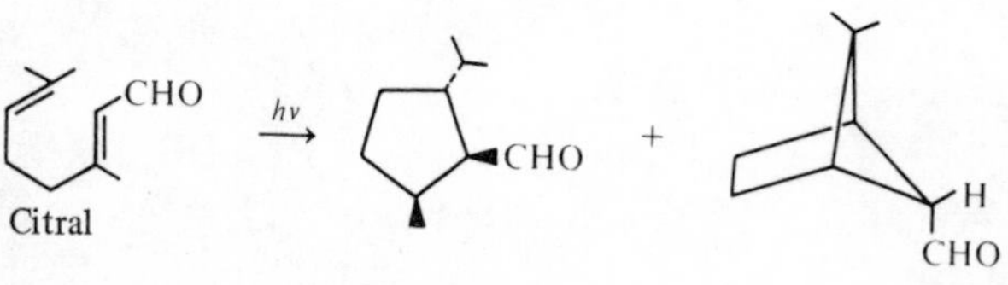

Citral

$\xrightarrow{h\nu}$

Carvone

Cyclic ketones give rearrangements which may or may not be accompanied by the reduction of the carbonyl group.

$\xrightarrow[CH_3OH]{h\nu}$ H C=C=O H ⇆

H H C=C=O $\xrightarrow{CH_3OH}$ H $CH_2CO_2CH_3$

The dimerization of cyclic α, β-unsaturated ketones may be illustrated by cyclopentenone:

O $\xrightarrow{h\nu}$ O O + O O

Open-chain α, β-unsaturated ketones give a *cis-trans* isomerizatio of the double bond; unless the isomeric product participates in another photochemical reaction, a *cis-trans* equilibrium is usuall attained.

The first stage in the liquid phase irradiation of *trans-α*-iononι a *trans-cis* isomerization, which is followed by the cleavage of th γ-proton, similarly to a type II Norrish reaction.

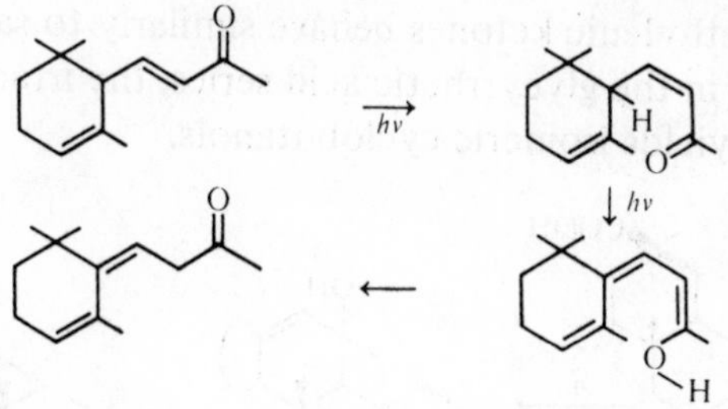

The same initial process is involved in the irradiation of *trans*-β-ionone, but the final products are a β, γ-ethylenic ketone and a pyrane isomer:

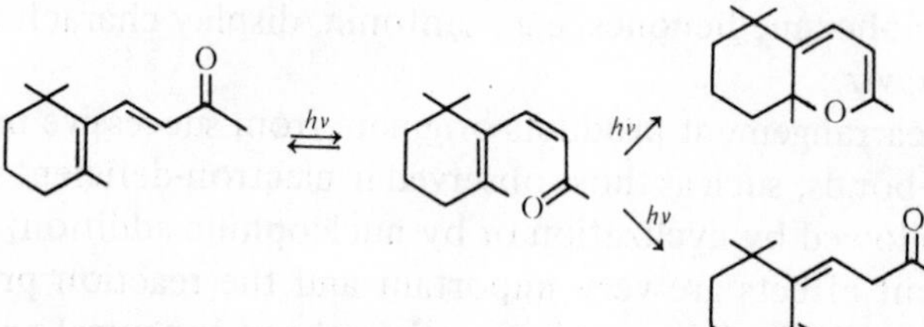

The photochemical isomerization of retinal is particularly important, since it is relevant to the visual process (Chapter 11).

In cyclic enones, *cis-trans* isomerization is usually sterically hindered, except when the ring size is sufficiently large; examples are 2-cyclooctenone and 2-cycloheptenone.

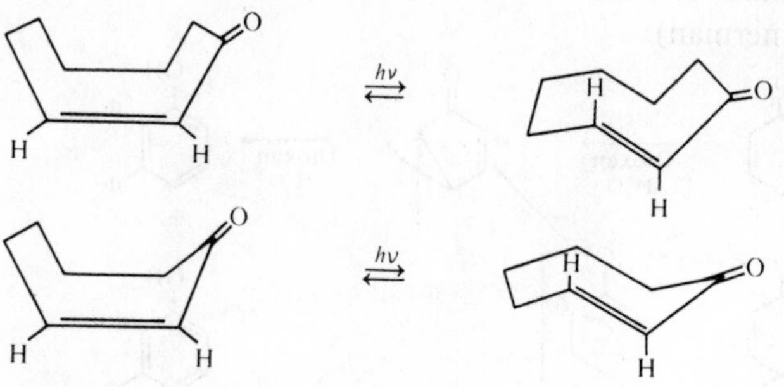

Certain α, β-ethylenic ketones behave similarly to saturated ketones. Thus, in the glycyrrhetic acid series, the irradiation of the C_{11}-ketone yields isomeric cyclobutanols.

COOH O 11 HO $\xrightarrow[\text{Dioxan}]{h\nu}$ OH HO + OH HO

Cyclic dienones /20–25/

2,5-Cyclohexanedienones, e.g., santonin, display characteristic properties, viz.:

a) the rearrangement products originate from successive migrations of σ-bonds, such as those observed in electron-deficient systems, and are followed by cyclization or by nucleophilic addition;

b) solvent effects are very important and the reaction products are different, depending on whether the solvent is neutral or acid, protonized or nonprotonized;

c) the presence of substituents affects the type of the rearrangement;

d) the same arrangement may sometimes take place if only one double bond is present, and not two.

The mechanistic treatment of these reactions may be found in a large number of different reviews and will not be dealt with here (H. Zimmerman).

O Φ Φ $\xrightarrow[\text{Dioxan, } H_2O]{h\nu}$ O Φ Φ $\xrightarrow[\text{Dioxan, } H_3O^{\oplus}]{h\nu}$ OH Φ Φ + OH Φ Φ + COOH Φ Φ

$h\nu$ $(n \rightarrow \pi^*)$ [O· Φ Φ ↕ O· Φ Φ → O· Φ Φ ↑ O$^{\ominus}$ $\oplus$ Φ Φ]

hν, HCO_2H, H_2O

hν, AcOH, Dioxan

favored in acid medium

favored in neutral medium

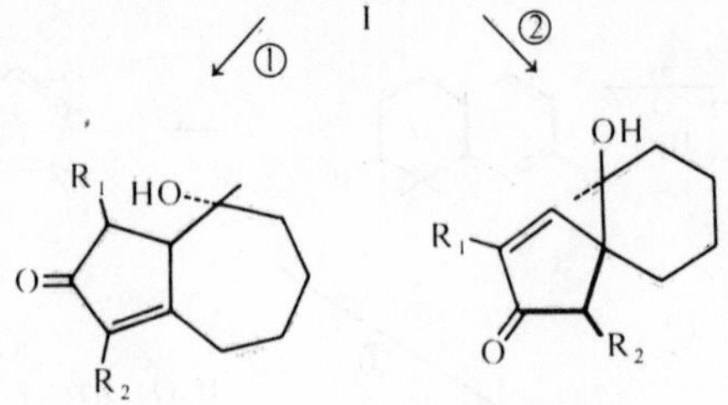

I

① ②

	①	②
$R_1{=}R_2{=}H$	19 %	16 %
$R_1{=}Me$; $R_2{=}H$	0	51 %
$R_1{=}H$; $R_2{=}Me$	50 %	0 %

The major product has the methyl group situated on the double bond.

The photochemistry of santonin has been fully elucidated /24, 25/.

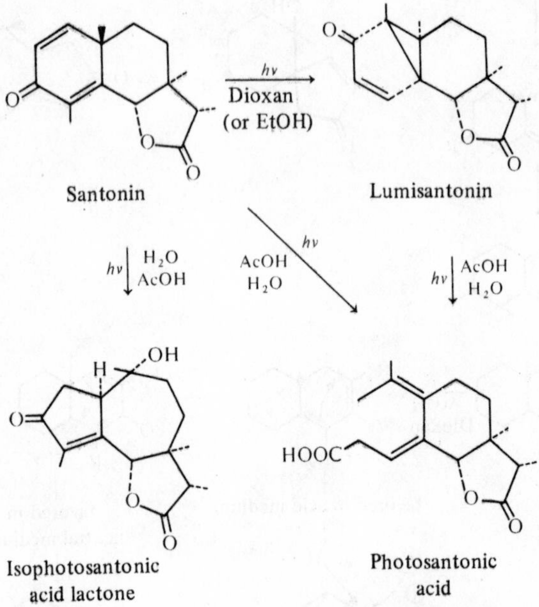

Santonin $\xrightarrow[\text{Dioxan (or EtOH)}]{h\nu}$ Lumisantonin

Santonin $\xrightarrow[H_2O,\ AcOH]{h\nu}$ Isophotosantonic acid lactone

Santonin $\xrightarrow[AcOH,\ H_2O]{h\nu}$ Photosantonic acid

Lumisantonin $\xrightarrow[AcOH,\ H_2O]{h\nu}$ Photosantonic acid

The mechanism of formation of lumisantonin and of isophotosantonic acid lactone may be schematically represented as follows:

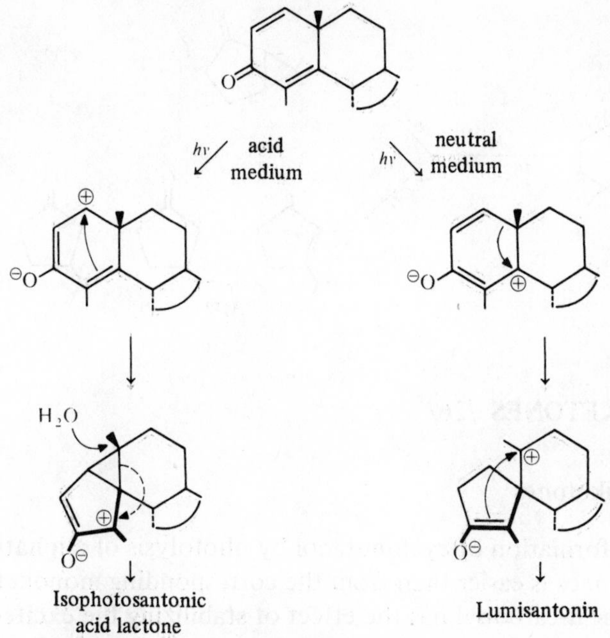

hν acid medium

hν neutral medium

H_2O

Isophotosantonic acid lactone

Lumisantonin

The mechanism of formation of photosantonic acid from lumisantonin can be schematically represented as follows:

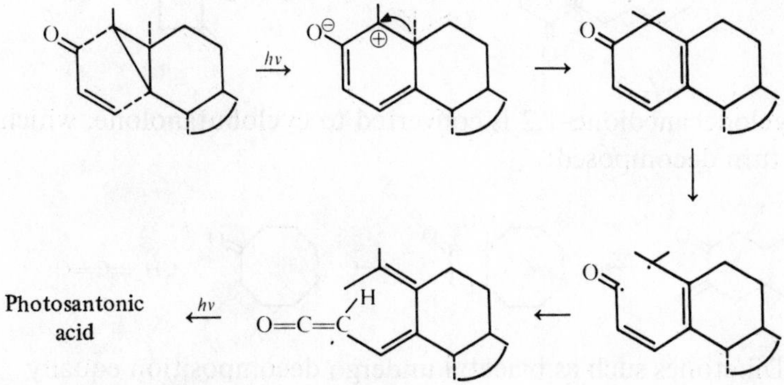

hν

Photosantonic acid

hν

Certain cyclic enones undergo photochemical rearrangements similar to those of dienones.

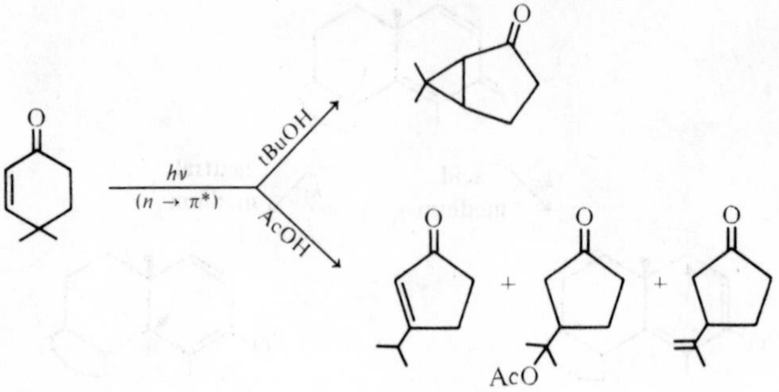

$h\nu$ $(n \rightarrow \pi^*)$ tBuOH AcOH

AcO

V. DIKETONES /26/

1. α-Diketones

The formation of cyclobutanol by photolysis of aliphatic α-diketones is easier than from the corresponding monoketones, since the α-carbonyl has the effect of stabilizing the excited state. A hydroxycyclobutanone is formed, as exemplified by 5,6-decanedione:

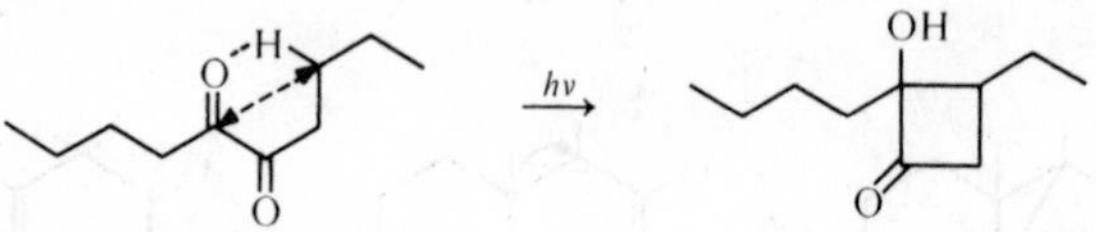

$h\nu$ OH

Cyclodecanedione-1,2 is converted to cyclobutanolone, which is in turn decomposed:

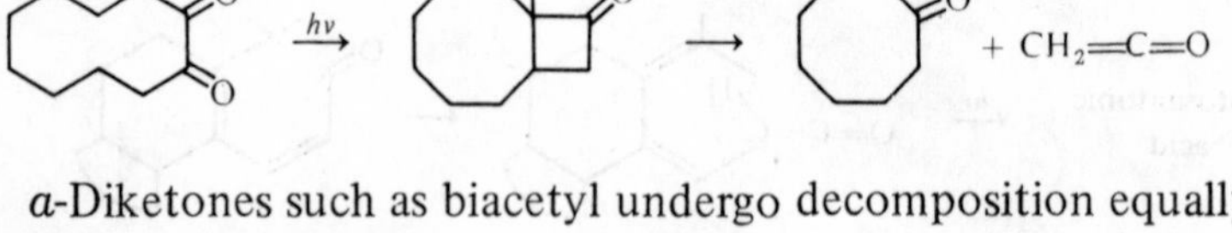

$h\nu$ OH $+ CH_2{=}C{=}O$

α-Diketones such as biacetyl undergo decomposition equally well in vapor phase and in solution:

$$CH_3{-}\overset{O}{\overset{\|}{C}}{-}\overset{O}{\overset{\|}{C}}{-}CH_3 \xrightarrow{h\nu} 2CO + CH_3{-}CH_3$$

2. β-Diketones

The principal reactions of β-diketones are decarbonylation and addition to olefins.

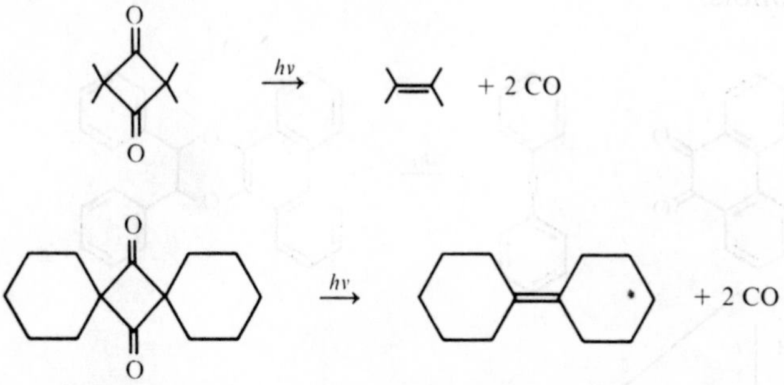

$\xrightarrow{h\nu}$ + 2 CO

$\xrightarrow{h\nu}$ + 2 CO

An intermediate cyclopropanone seems to be involved in these decarbonylations.

β-Diketones add onto olefins in their enolic form:

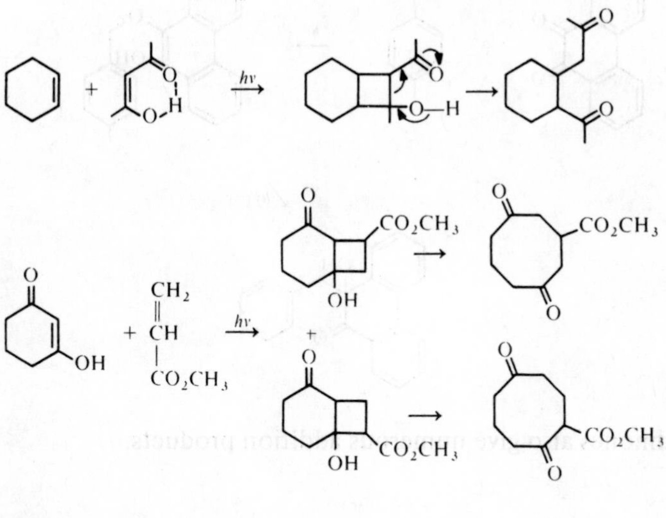

$\xrightarrow{h\nu}$ →

CO_2CH_3 → CO_2CH_3

OH

+ $CH_2 = CH - CO_2CH_3$ $\xrightarrow{h\nu}$ +

OH

→ CO_2CH_3

OH CO_2CH_3

3. Quinones

o-Quinones add onto double bonds to form dioxenes, onto aldehydes to form esters, and onto hydrocarbons and ethers to form alcohols.

R—C—H ‖ O hν

hν

O—C—R ‖ O OH

O O OH

hν

O OH

hν

OH O

p-Quinones also give numerous addition products.

CHO
Cl Cl Cl Cl O O O hv Cl Cl Cl Cl OH
HO O—CH_2 Cl Cl Cl Cl

O hv O O
Φ Φ hv O Φ Φ O Φ O Φ O
O hv O O O
O O hv O O O O
CHO hv OH O OH

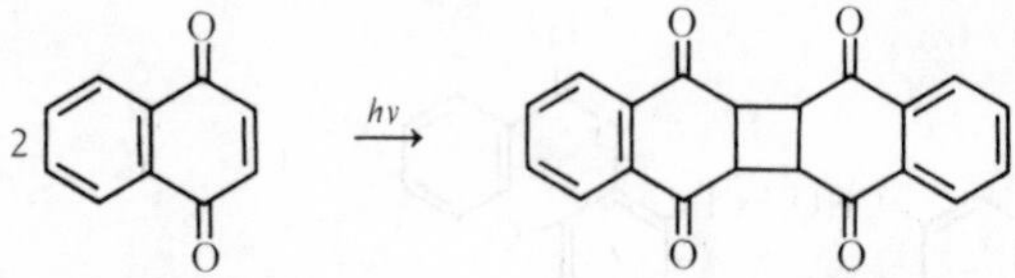

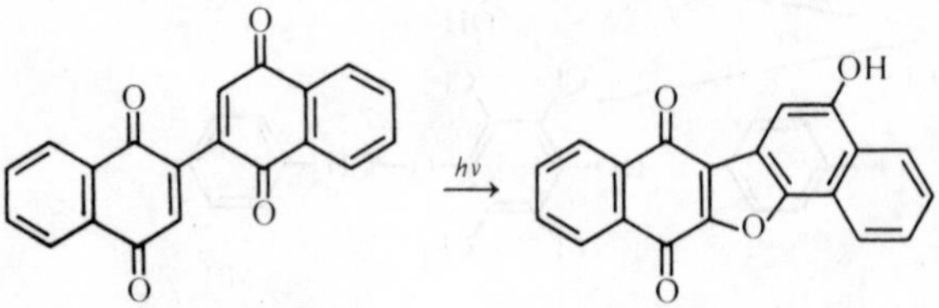

VI. α, β-EPOXY ALDEHYDES AND KETONES /27–29/

Vapor phase irradiation of glycidyl aldehyde mainly results in the decomposition of the molecule.

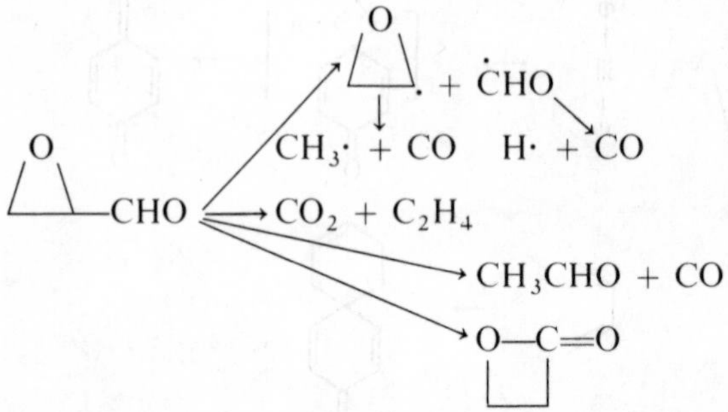

Photolysis of α, β-epoxy ketones with an alkyl group bound to the carbonyl initially gives β-diketones; those with a phenyl group and with a hydrogen atom in position γ to the carbonyl yield α-hydroxy ketones.

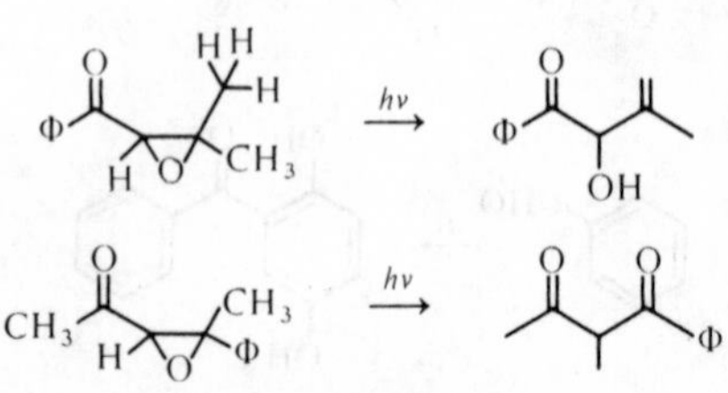

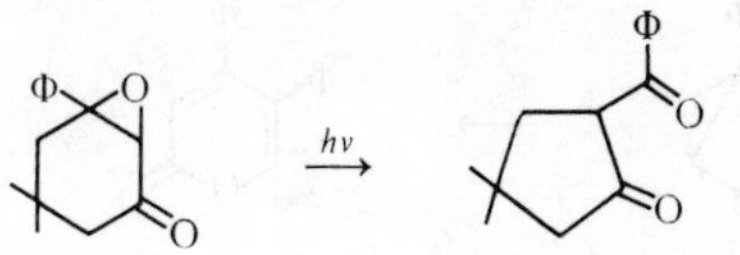

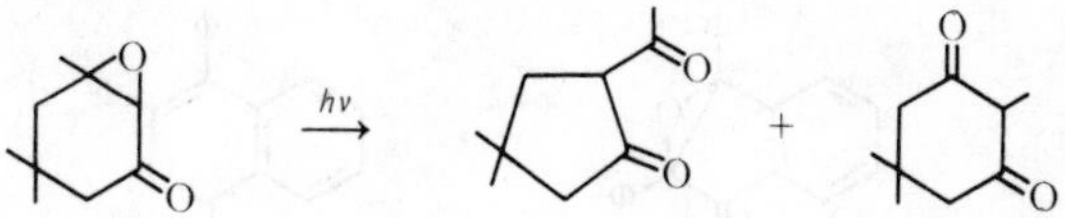

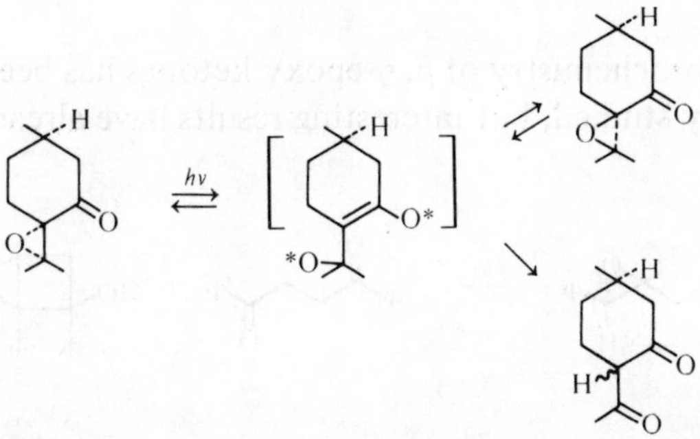

Φ O hν Φ O O O

O O hν O O + O O

H hν H O* *O H O O H H O O

Jeger et al., and also other workers, studied numerous photolyses of α, β-epoxy ketosterols. Jeger noted that the 10α-configuration is retained during the rearrangements of α- or β-epoxides of 10α-testosterone.

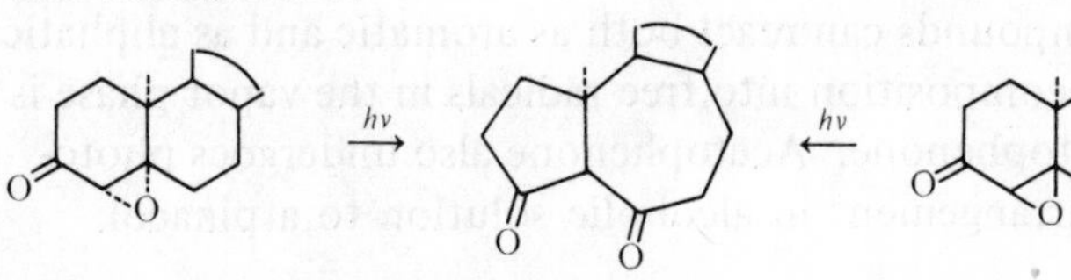

O O hν O O hν O O

γ, δ-Epoxy-α, β-unsaturated ketones also undergo rearrangements:

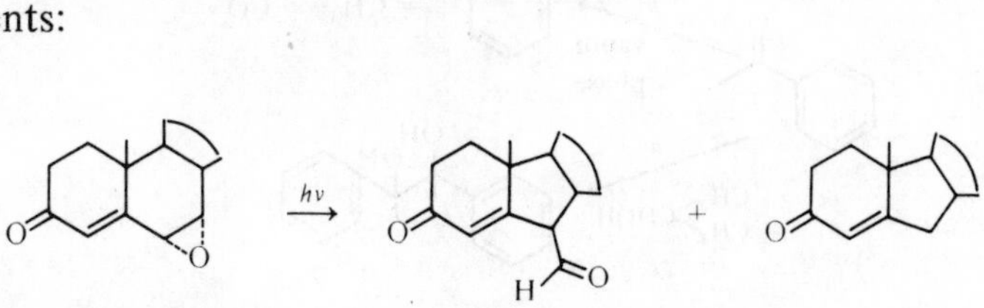

O O hν O H O + O

as do α, β-epoxy-α′, β′-unsaturated ketones:

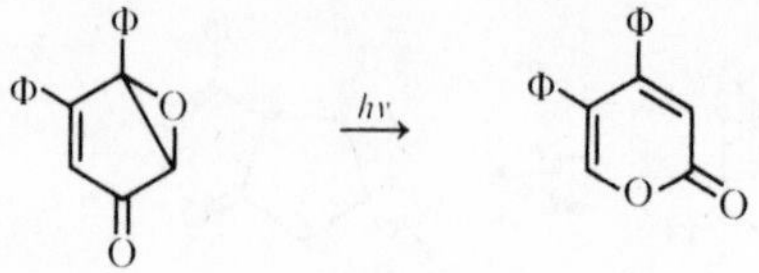

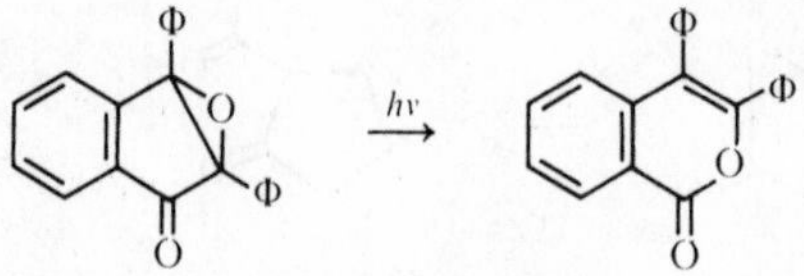

The photochemistry of β, γ-epoxy ketones has been much less thoroughly studied, but interesting results have already been obtained:

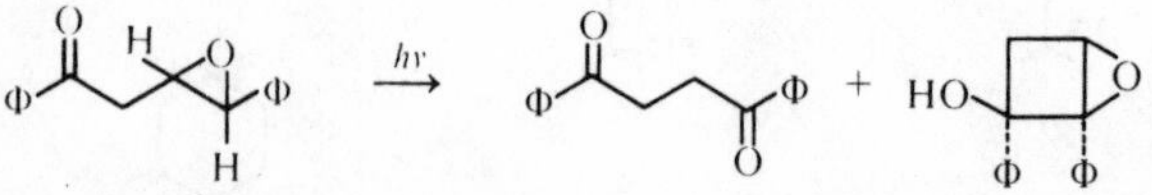

VII. AROMATIC ALDEHYDES AND KETONES

1. Alkyl aryl ketones

These compounds can react both as aromatic and as aliphatic ketones. Decomposition into free radicals in the vapor phase is given by acetophenone. Acetophenone also undergoes photochemical rearrangement in alcoholic solution to a pinacol.

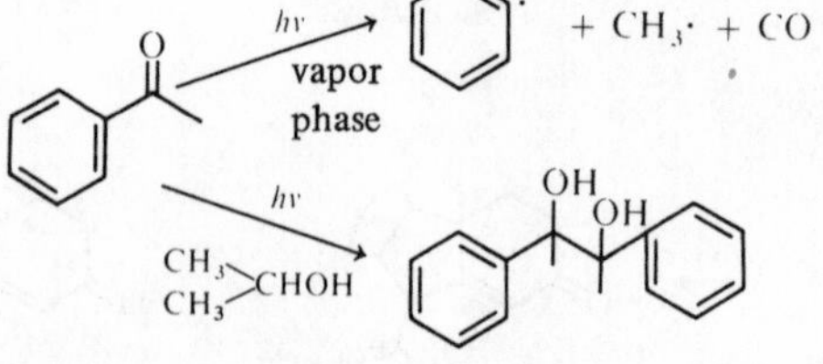

$h\nu$ vapor phase → $+ CH_3\cdot + CO$

$h\nu$ $\frac{CH_3}{CH_3}>CHOH$ → OH OH

Ketones with a γ-hydrogen, such as butyrophenone, mainly give type II decompositions. Pitts et al. made a systematic study of the quantum yields of this reaction for various substituted butyrophenones and showed that only ketones with ($n \rightarrow \pi^*$) as their lowest triplet state were reactive.

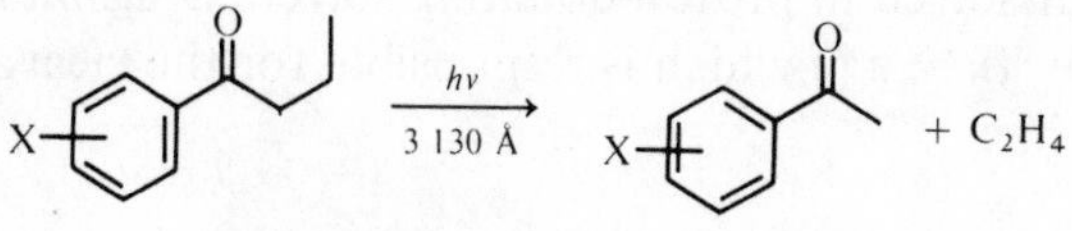

$$\xrightarrow[3\,130\ \text{Å}]{h\nu} \quad + \ C_2H_4$$

o-Methylbutyrophenone does not give a type II reaction, even though $^3(n \rightarrow \pi^*)$ is the lowest triplet state; when the enolic form is stabilized at a low temperature, a reversible photoenolization is observed; it is a photochromic reaction.

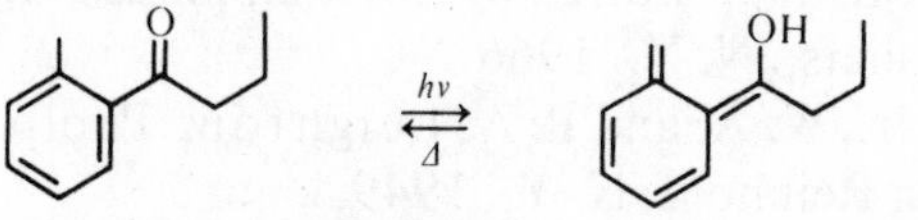

As regards halogenated butyrophenones (X = Br or I), elimination of the halogen is favored rather than type II decomposition, whereas type II decomposition is the only reaction which is observed if X = F or Cl.

2. Benzaldehyde

When irradiated in alcoholic solution, benzaldehyde yields hydrobenzoin.

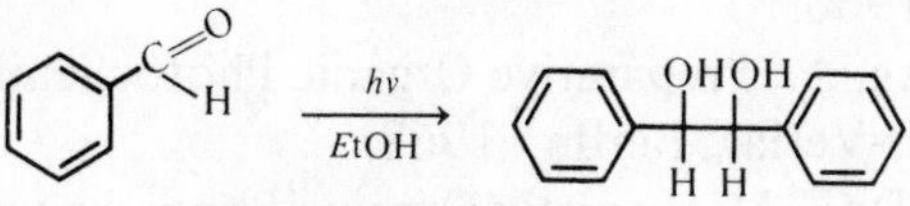

$$\xrightarrow[EtOH]{h\nu}$$

Addition to olefinic double bonds yields oxetanes, while the addition to triple bonds yields the unstable unsaturated oxetanes (oxetes).

o-Nitrobenzaldehyde undergoes rearrangement to *o*-nitrosobenzoic acid both in solution and in the solid state. This photoisomerization has been utilized in actinometry (cf. Chapters 4 and 14).

3. Benzophenones /30/

Pinacol formation is the main photochemical reaction of benzophenones dissolved in proton-donating solvents; again, it is the triplet state $^3(n \rightarrow \pi^*)$ which is responsible for the cleavage of the hydrogen.

Bibliography

1. Pitts, Jr., J. N. and J. K. S. Wan. The Chemistry of the Carbonyl Group. Edited by S. Patai, p.823. Interscience Publishers, N. Y. 1966.
2. Noyes, Jr., W. A. and P. A. Leighton. Photochemistry of Gases, Reinhold, N. Y. 1949.
3. Davis, Jr., W. – Chem. Rev., **40**:201. 1947.
4. Rollefson, G. K. and M. Burton. Photochemistry and the Mechanisms of Chemical Reactions. Englewood Cliffs, Prentice Hall, N.J. 1939.
5. Pitts, Jr., J. N. – J. Chem. Educ., **34**:112. 1957.
6. Noyes, Jr., W. A., G. B. Porter, and J. E. Jolley. – Chem. Rev., **56**:49. 1956.
7. Yang, N. C. Reactivity of the Photoexcited Organic Molecules, p.145. Interscience Publishers, N. Y. 1967.
8. Kan, R. O. Organic Photochemistry. McGraw-Hill Company, N. Y. 1966.

   Schönberg, A. Preparative Organic Photochemistry. Springer-Verlag, Berlin. 1968.

   Neckers, D.C. Mechanistic Organic Photochemistry. Reinhold, N. Y. 1967.

   Calvert, J. G. and J. N. Pitts, Jr. Photochemistry, pp.366–685. John Wiley and Sons, Inc., N. Y. 1966.

   Turro, N. J. Molecular Photochemistry. W. A. Benjamin, Inc., N. Y. 1965.
9. Chapman, O. L. Advances in Photochemistry. Edited by W. A. Noyes Jr., G. S. Hammond, and J. N. Pitts Jr., 1:323. Interscience Publishers, N. Y. 1963.

10. Chapman, O. L. and G. Lenz. Organic Photochemistry. Edited by O. L. Chapman, **1**: 283. M. Dekker, N. Y. 1967.
11. Yang, N. C. – Pure Appl. Chem., **9**:591. 1964.
12. Niclause, M., J. Lemaire, and M. Letort. Advances in Photochemistry. Edited by W. A. Noyes, Jr., G. S. Hammond, and J. N. Pitts, Jr., **4**: 25. Interscience Publishers, N. Y. 1966.
13. Srinivasan, R. Ibid., **1**:83. 1963.
14. Quinkert, G. – Pure Appl. Chem., **9**:607. 1964.
15. Yates, P. – Ibid., **16**:93. 1968.
16. Cookson, R. C. – Ibid., **9**:575. 1964.
17. Eaton, P. E. – Acc. Chem. Res., **1**:50. 1968.
18. Pasto, D. J. Organic Photochemistry. Edited by O. L. Chapman, **1**:155, M. Dekker, N. Y. 1967.
19. Mousseron, M. and J. P. Chabaud. – Bull.Soc.Chim.France, p.239. 1969.
20. Kropp, P. J. Organic Photochemistry. Edited by O.L. Chapman, **1**:1. M. Dekker, N. Y. 1967.
21. Schaffner, K. Advances in Photochemistry. Edited by W. A. Noyes, Jr., G. S. Hammond, and J. N. Pitts, Jr., **4**:81. Interscience Publishers, N. Y. 1966.
22. Zimmerman, H. Ibid., **1**:183. 1963.
23. Zimmerman, H. – Pure Appl. Chem., **9**:493. 1964.
24. Barton, D. H. R., P. de Mayo, and M. Shafiq. – J. Chem. Soc., 929. 1957.
25. Simonsen, J. and D. H. R. Barton. The Terpenes, **3**:292. Cambridge University Press. 1952.
26. De Mayo, P. – Pure Appl. Chem., **9**:597. 1964.
27. Schaffner, K. – Ibid., **16**:75. 1968.
28. Padwa, A. Organic Photochemistry. Edited by O.L. Chapman, **1**:91. M. Dekker, N. Y. 1967.
29. Jeger, O., K. Schaffner, and H. Wehrli. – Pure Appl. Chem., **9**:555. 1964.
30. Porter, G. and P. Suppan. – Ibid., **9**:499. 1964.

Chapter 8

PHOTOSENSITIZED OXIDATIONS

Reactions of oxidation by molecular oxygen in the presence of light have been known for a long time (Windaus, 1928). The various types of reactions between oxygen and the substrate can be classified in accordance with the nature of the electronic states of oxygen and of the substrate, as follows:

a) combination of oxygen and substrate when both are in the ground state:

$$A + O_2 \rightarrow AO_2$$

b) attack by molecular oxygen in the ground state on electronically excited substrate without a sensitizer:

$$A \xrightarrow{h\nu} A^* \xrightarrow{O_2} AO_2$$

c) autoxidation, by molecular oxygen in the ground state, of a radical formed by the cleavage of a hydrogen atom off the substrate by an excited sensitizer:

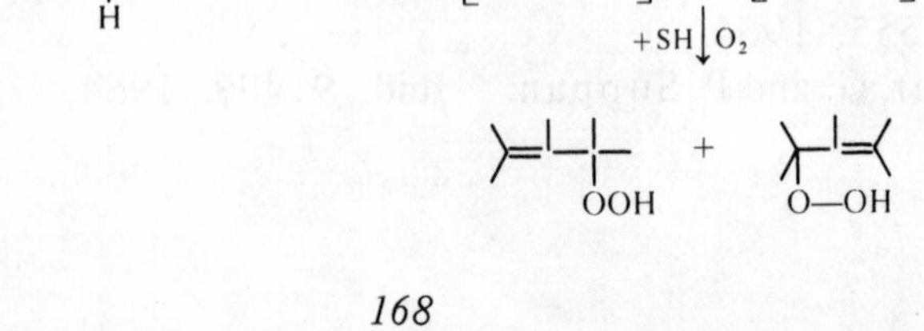

H $+ S_3^* \xrightarrow[-SH]{}$ [ ] $\leftrightarrow$ [ ]

$+SH \downarrow O_2$

OOH + O—OH

This process will not be discussed in this chapter;

d) attack by excited molecular oxygen on the substrate in the ground state. This is a very important type of reaction, which will be discussed in some detail.

Owing to the fact that the oxygen molecule has very special electronic features, we must begin by a detailed study of this molecule.

I. ELECTRONIC STATES OF OXYGEN

Elements of the second period (lithium to fluorine) have four bond-forming orbitals: $2s$, $2p_x$, $2p_y$, and $2p_z$. We shall consider a diatomic molecule A_2, with the straight line A–A as the Z-axis, and construct the molecular orbitals. The two atomic orbitals $2s$ give two nonbonding orbitals at the two ends of the molecule. The two atomic orbitals $2p_z$ combine to form the two orbitals σ and σ^*; the two $2p_x$ orbitals form the two orbitals π_x and π_x^*, and the two $2p_y$ orbitals form the two orbitals π_y and π_y^*. For reasons of symmetry, the energies of π_x and π_y are equal; the same applies to π_x^* and π_y^*. In a nitrogen molecule N_2, 10 electrons must be placed in the bonding orbitals σ, π_x, π_y and in the two nonbonding orbitals,

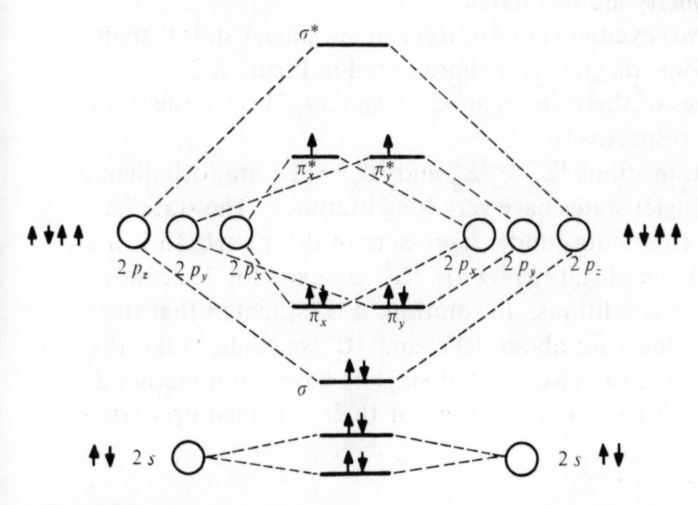

FIGURE 8.1

each orbital being doubly occupied. The oxygen molecule O_2 has 12 electrons: after all the bonding and nonbonding orbitals have been filled, the two remaining electrons must be placed in the lowest energy antibonding orbitals π_x^* and π_y^*.

In the ground state, one of these electrons occupies the π_x^* orbital while the other occupies the π_y^* orbital; their spins are parallel, which results in the minimum electrostatic interaction and yields a triplet ground state $^3\Sigma_g^-$ (Figure 8.1).

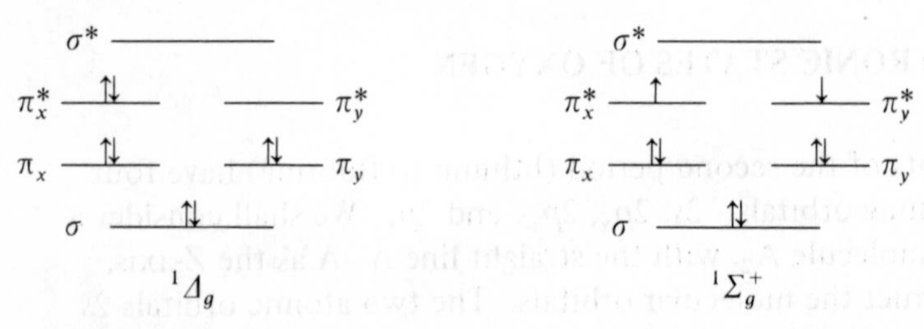

FIGURE 8.2

In the ground state oxygen has a triple bond (one σ-bond and two π-bonds) and two half-antibonds; this is equivalent to a double bond. Since its ground state is a triplet, nonexcited oxygen is paramagnetic and, in addition, processes involving a change in the multiplicity are facilitated.

The first two excited states of oxygen are singlet states; their partial electronic diagrams are represented in Figure 8.2.

The energies of these states are 22.5 and 37.5 kcal·mole^{-1} for $^1\Delta_g$ and $^1\Sigma_g^+$, respectively.

Since the transitions $^1\Sigma_g^+ \rightarrow {}^3\Sigma_g^-$ and $^1\Delta_g \rightarrow {}^3\Sigma_g^-$ are forbidden, the excited singlet states have very long lifetimes. The state $^1\Delta_g$ can exist for 45 minutes under a pressure of 0.1 mm Hg (it is the longest-lived excited state known); $^1\Sigma_g^+$ can exist for 7 seconds under the same conditions. In solution, it is estimated that the respective lifetimes are about 10^{-3} and 10^{-8} seconds. Like the ground state $^3\Sigma_g^-$, the two excited singlets have a double bond structure, which accounts for some of their chemical properties.

II. MECHANISM OF PHOTOSENSITIZED OXIDATIONS

Two mechanisms have been proposed for this type of oxidation: Schönberg's "moloxide" mechanism and Kautsky's singlet oxygen mechanism.

1. "Moloxide" mechanism

This mechanism, which was originally proposed by Schönberg, was subsequently taken up by Schenck, Livingston and Oester, and by other scientists. It postulates that a complex (moloxide) between the sensitizer S and oxygen is the intermediate compound, which then reacts with the substrate A:

$$^{1}S_0 + h\nu \rightarrow {}^{1}S_1$$

$$^{1}S_1 \rightsquigarrow {}^{3}S_1$$

$$^{3}S_1 + {}^{3}(O_2)_0 \rightarrow (S \text{ --- } O_2)^*$$

$$A + (S \text{ --- } O_2)^* \rightarrow AO_2 + {}^{1}S_0 \,.$$

The "moloxide" is an excited charge transfer complex, the multiplicity of which can be 1, 3, or 5.

2. Singlet oxygen mechanism

In this mechanism, the initial stages are the same as in the "moloxide" mechanism, but the substrate-attacking species is different, viz., it is the singlet oxygen formed by the transfer of the electronic energy of the sensitizer. This energy transfer occurs via a charge transfer complex (Chapter 3):

$$^{1}S_0 + h\nu \rightarrow {}^{1}S_1$$

$$^{1}S_1 \rightsquigarrow {}^{3}S_1$$

$$^{3}S_1 + {}^{3}(O_2)_0 \rightarrow (S \text{ --- } O_2)^*$$

$$(S \text{ --- } O_2)^* \rightarrow {}^1S_0 + {}^1(O_2)_1$$

$${}^1(O_2)_1 + A \rightarrow AO_2\ .$$

Contrary to the "moloxide" mechanism, this mechanism provides an explanation for the photooxidations in solid media: the distance between the substrate and the sensitizer is such that the only factor which can account for the observed reaction is the diffusion of the oxidizing species from the sensitizer to the substrate (experiments of Kautsky and of Bourdon-Schnuriger). This oxidizing species, which is capable of migrating through the solid medium, would be molecular oxygen in the excited singlet state ${}^1\Delta_g$ or ${}^1\Sigma_g^+$.

It was shown by Foote and Wexler that singlet oxygen, prepared chemically or electrically, gave the same results as photosensitized oxidation, thus confirming the mechanism proposed by Kautsky.

3. Preparation of excited molecular oxygen

These methods were studied by Foote and Wexler, and also by Corey.

A) *Direct photochemical excitation*

Even though the transitions

$${}^3\Sigma_g^- \xrightarrow{h\nu} {}^1\Delta_g \quad \text{or} \quad {}^3\Sigma_g^- \xrightarrow{h\nu} {}^1\Sigma_g^+$$

are highly improbable, they cannot be neglected as far as the irradiation of atmospheric oxygen by sunlight is concerned.

Singlet oxygen can also be prepared by the action of very hard UV radiation.

B) *Photolysis of ozone*

$$O_3 \xrightarrow{\lambda > 5\,900\ \text{Å}} {}^3(O_3)^* \longrightarrow O_2({}^3\Sigma_g^-) + O({}^3P)$$

$$O_3 \xrightarrow{\lambda < 5\,900\ \text{Å}} {}^3(O_3)^* \longrightarrow O_2({}^1\Delta_g) + O({}^3P)$$

$$O_3 \xrightarrow{\lambda \sim 300\ \text{Å}} {}^1(O_3)^* \longrightarrow O_2({}^1\Delta_g + {}^1\Sigma_g^+) + O({}^1D)\ .$$

C) *Chemical formation in the atmosphere*

According to Pitts et al., two reactions which take place in the atmosphere also generate excited oxygen, and are thus relevant to the problem of air pollution:

$$NO + O_3 \rightarrow NO_2 + O_2(^1\Delta_g)$$
$$NO_2 + O \rightarrow NO + O_2(^1\Delta_g)\,.$$

However, the formation of singlet oxygen in these reactions has not yet been conclusively demonstrated.

D) *Chemical formation in the liquid phase*

Singlet oxygen can be produced in a 10% yield by decomposing hydrogen peroxide with reagents such as hypochlorite, bromine, etc., which have a positive halogen:

$$H_2O_2 + ClO^- \rightarrow H_2O + Cl^- + O_2(^1\Delta_g)\,.$$

Excited oxygen is also liberated by the action of hydrogen peroxide on a nitrile:

$$R{-}CN + HO{-}O^- \longrightarrow R{-}C(O{-}OH){=}N^- \xrightarrow{H_2O} R{-}C(O{-}OH){=}NH + OH^-$$

$$R{-}C(O{-}OH){=}\ddot{N}H \quad H{-}O{-}O{-}H \longrightarrow R{-}C({=}O){-}NH_2 + H_2O + {}^1(O_2)$$

Finally, excited oxygen can also be produced by thermal decomposition of a peroxide: thus, when substituted anthracene peroxides are heated, the parent hydrocarbon is reconstituted and excited molecular oxygen is evolved:

$\xrightarrow{\Delta}$ + O_2^*

E) *Microwave excitation*

The various techniques utilized by Corey make it possible to excite molecular oxygen by microwave discharge in a tube or in a resonant cavity; the frequencies used were 5, 6, 7, and 2450 MHz. When 100–250 W generators and 0.5–5 mm Hg oxygen pressures were used, 10–20% yield of $^1\Delta_g$ oxygen were obtained.

The singlet $^1\Sigma_g^+$ can be obtained by the dismutation

$$^1\Delta_g + {}^1\Delta_g \rightarrow {}^3\Sigma_g^- + {}^1\Sigma_g^+ .$$

III. PHOTOSENSITIZED OXIDATION OF 1,3-DIENES

1. Cyclohexadienes

Photosensitized oxidation of cyclohexadienes gives high yields of endocyclic peroxides.

It was shown by Foote and Wexler that the same endoperoxide was formed by the action of chemically produced ($NaOCl + H_2O_2$) singlet oxygen in the absence of light.

Various examples of endoperoxide formation by photosensitized oxidation are known:

$\xrightarrow[\text{Sens.}]{O_2/h\nu}$

$\xrightarrow[\text{Sens.}]{O_2/h\nu}$

The stability of 1,4-endoperoxides is usually poor, and these compounds readily undergo decomposition and rearrangement. However, some of them are very stable; the natural peroxide ascaridol is an example. Its synthesis from *α*-terpinene was carried out by Schenck and Ziegler.

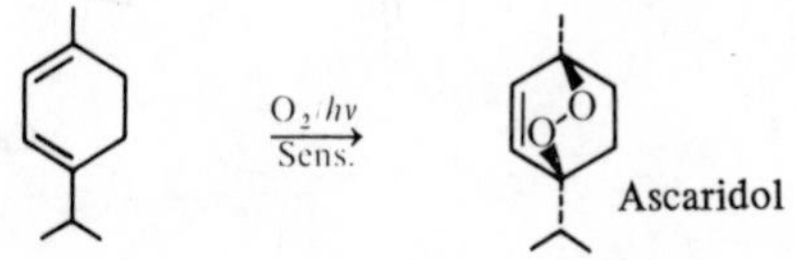

$\xrightarrow[\text{Sens.}]{O_2/h\nu}$ Ascaridol

The singlet oxygen attack is directed by the substituents of the substrate; thus, photooxidation of *levo*-pimaric acid yields the *α*-endoperoxide only.

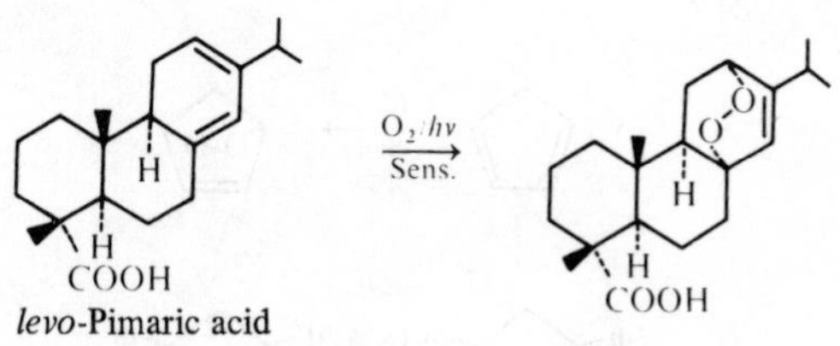

levo-Pimaric acid $\xrightarrow[\text{Sens.}]{O_2/h\nu}$

Many examples of stereospecific attacks on cyclohexadiene compounds are known.

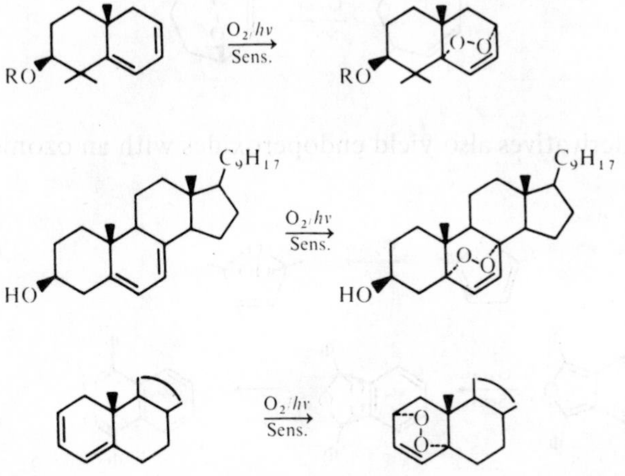

During the Schenck synthesis of cantharidin, the diene is preferentially attacked on one side only.

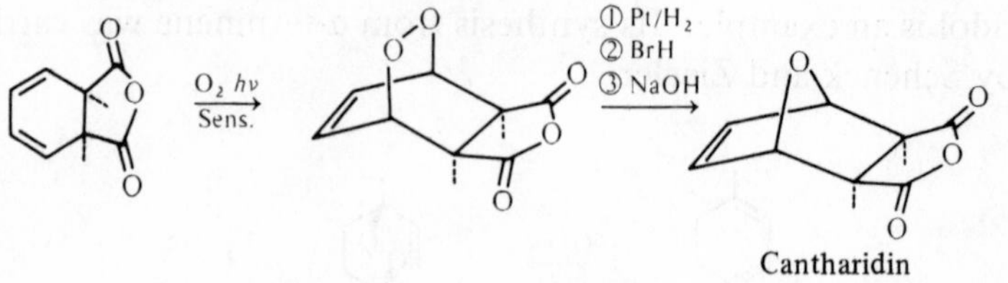

O_2 $h\nu$ / Sens. → ① Pt/H_2 ② BrH ③ NaOH →

Cantharidin

2. Various homocyclic dienes

Photosensitized oxidations of cyclopentadienes and of 1,3-cycloheptadienes are known. The endoperoxide of cyclopentadiene is very unstable, but can be stabilized by phenyl substituents:

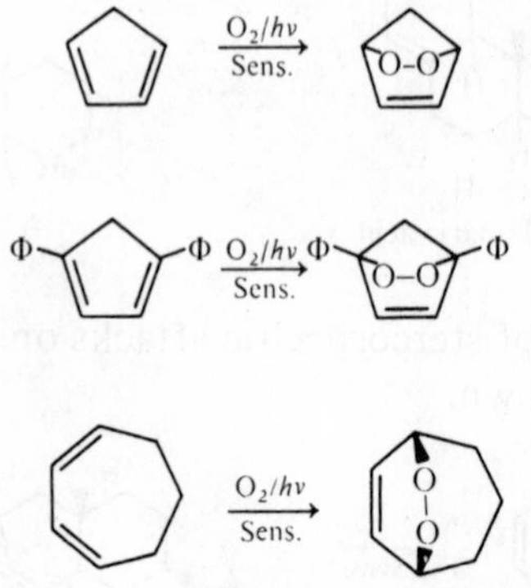

$O_2/h\nu$ / Sens.

$O_2/h\nu$ / Sens.

$O_2/h\nu$ / Sens.

Furan derivatives also yield endoperoxides with an ozonide structure:

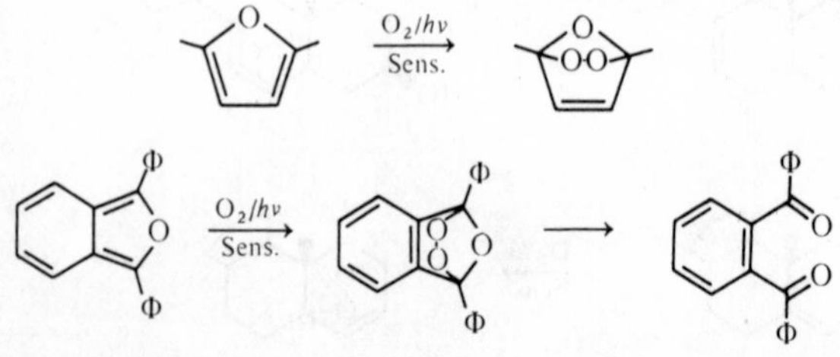

$O_2/h\nu$ / Sens.

$O_2/h\nu$ / Sens.

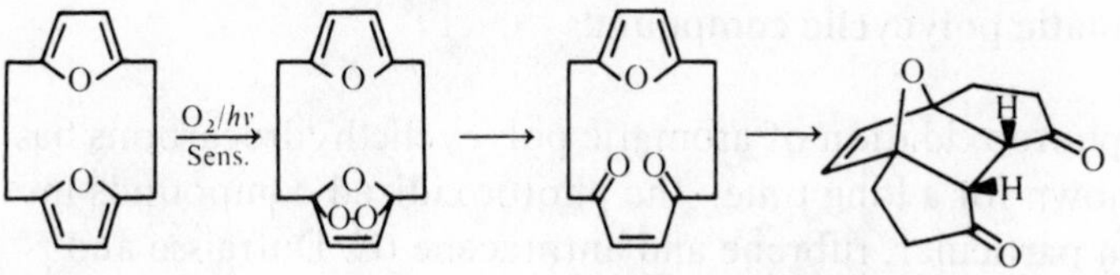

Pyrroles in which the nitrogen atom is not substituted form hydroperoxides rather than endoperoxides. If the nitrogen is substituted and the formation of a hydroperoxide is thus prevented, an endoperoxide is formed; it is unstable, and is rapidly rearranged in acid medium:

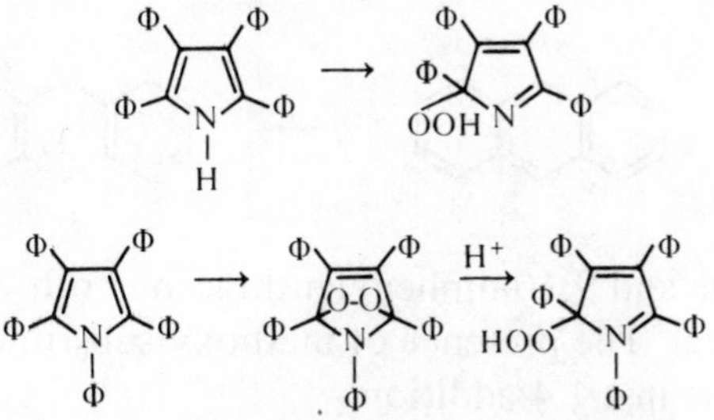

Photooxidation of oxazoles yields unstable endoperoxides; this reaction has been utilized in the preparation of certain ω-cyano acids.

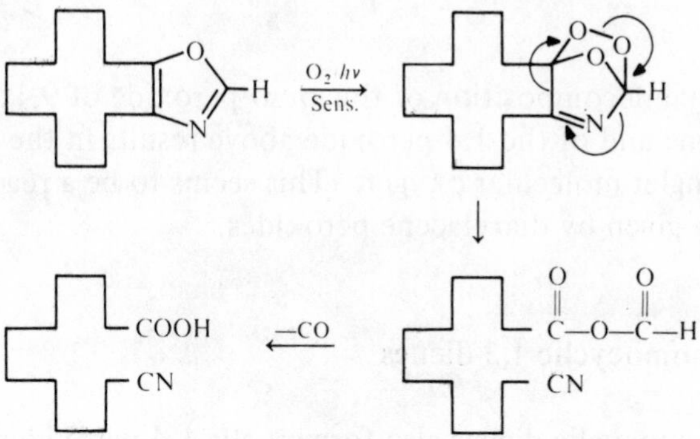

3. Aromatic polycyclic compounds

The photooxidation of aromatic polycyclic hydrocarbons has been known for a long time; the photooxidized compounds include, in particular, rubrene and anthracene (C. Dufraisse and J. Rigaudy).

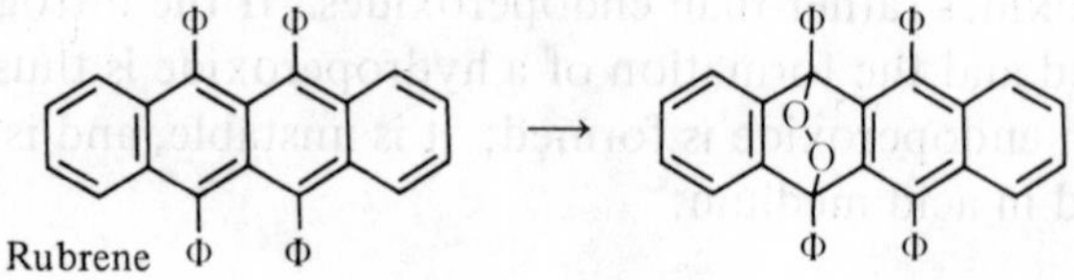

Anthracene and 9,10-diphenylanthracene both yield a 9,10-*meso*-peroxide. The presence of methoxy substituents in positions 1 and 4 results in a 1,4-addition:

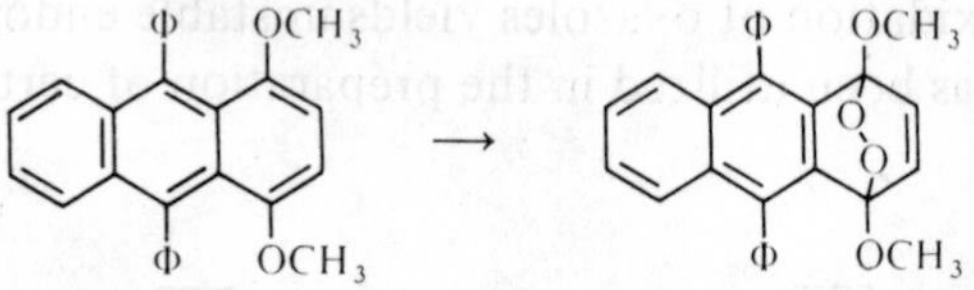

Thermal decomposition of the *meso*-peroxide of 9,10-diphenylanthracene and of the 1,4-peroxide above results in the liberation of the singlet molecular oxygen. This seems to be a reaction generally given by diarylacene peroxides.

4. Nonhomocyclic 1,3-dienes

Nonhomocyclic dienes also form cyclic 1,4-peroxides. Except for a few recent studies, this type of photooxidation has not yet been studied to any extent. These include Barton's photooxidation of a heteroring diene in the ergosterol series; the unstable peroxide cannot be isolated and undergoes rearrangement:

C_9H_{17} AcO $\xrightarrow[\text{Sens.}]{O_2/h\nu}$ [C_9H_{17} AcO O O]

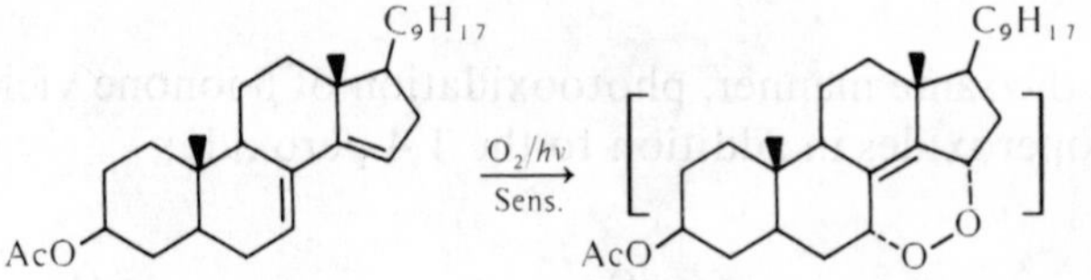

The photooxidation of 1,1′-dicyclohexene yields the *cis*-peroxide only, in accordance with K. Gollnick's concerted mechanism of oxygen attack:

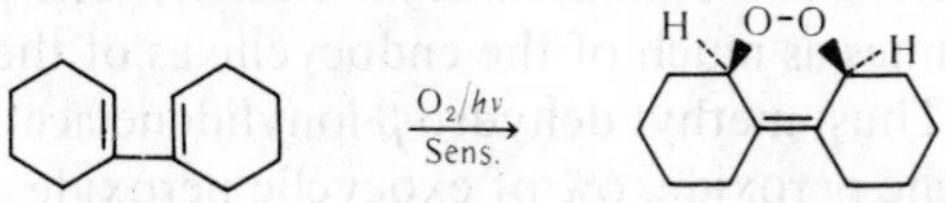

$\xrightarrow[\text{Sens.}]{O_2/h\nu}$ H O–O H

The peroxide of β-ionone is unstable and undergoes rearrangement to the hemiacetal; the peroxide of retinal is more stable and could be isolated (M. Mousseron):

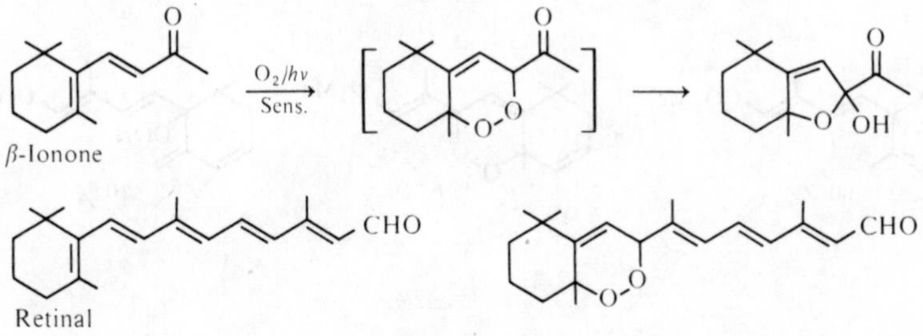

O $\xrightarrow[\text{Sens.}]{O_2/h\nu}$ [O O O] ⟶ O O OH

β-Ionone

CHO CHO O O

Retinal

It will be noted that nonhomocyclic dienes often tend to behave like compounds with isolated double bonds, and hydroperoxides as well as 1,4-peroxides are usually formed:

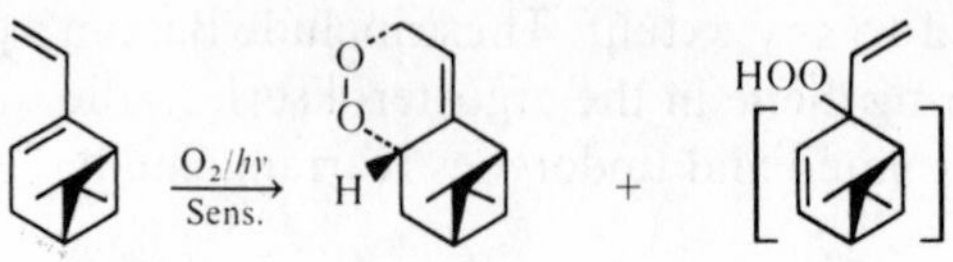

$\xrightarrow[\text{Sens.}]{O_2/h\nu}$ H + HOO

In the same manner, photooxidation of β-ionone yields two hydroperoxides in addition to the 1,4-peroxide:

O OH O OH

Dehydro-β-ionone compounds yield two types of dienes: a cyclohexadiene and a nonhomocyclic diene. The photooxidation yields 10 times as much of the endocyclic as of the exocyclic peroxide. Thus, methyl dehydro-β-ionylidene acetate yields 60% of endocyclic peroxide, 6% of exocyclic peroxide and 10% of hydroperoxide:

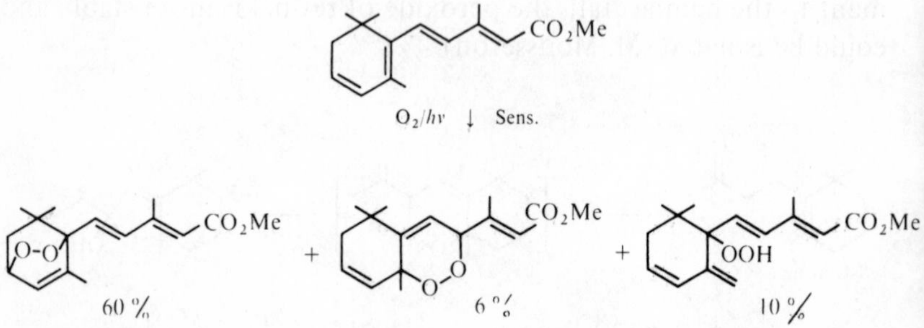

CO_2Me

$O_2/h\nu$ ↓ Sens.

CO_2Me + CO_2Me + CO_2Me OOH

60 % 6 % 10 %

IV. PHOTOSENSITIZED OXIDATION OF ISOLATED DOUBLE BONDS

The photosensitized oxidation of an olefin usually yields allylic hydroperoxides which have undergone migration of the double bond. The oxygen adds onto the carbons of the double bond and detaches a hydrogen atom from the allylic position to form an allylic hydroperoxide (Schenck reaction).

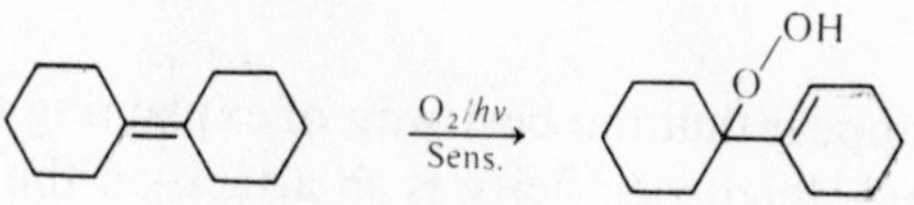

The reactivity of tetrasubstituted double bonds is comparable to that of cyclohexadienes. Trisubstituted, disubstituted, and monosubstituted double bonds react at diminishing rates, in that order.

The hydroperoxides are most often highly unstable and difficult to isolate. As a rule, it is preferable to prepare the corresponding alcohols directly by reducing with sodium sulfite or triphenylphosphine.

In the sterol series, the hydrogen atom is detached from a quasi-axial position which is *cis* to the C–O bond formed. Since the attack on the β-face is sterically hindered, the oxygen will preferentially attack the α-face. This mechanism has been confirmed by a 7β-photooxidation of the deuterated cholesterol:

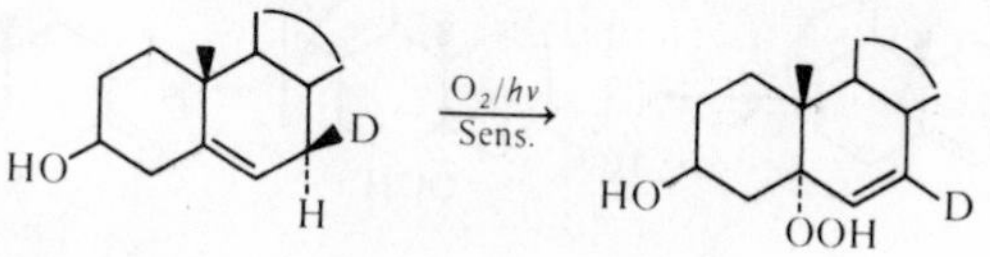

5α-Hydroxy-Δ^6-cholesterol, which has no α-*cis*-axial hydrogen atom, does not become photooxidized.*

* Photooxidation of isolated double bonds without an α-hydrogen has been recently observed. The products include two carbonyl compounds; the postulated intermediate compound is a ring peroxide with four elements identical with those involved in chemiluminescence processes (p.205).

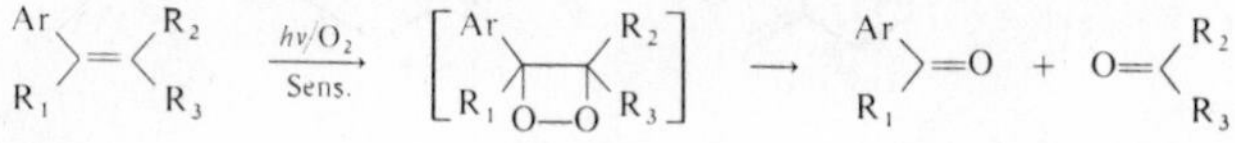

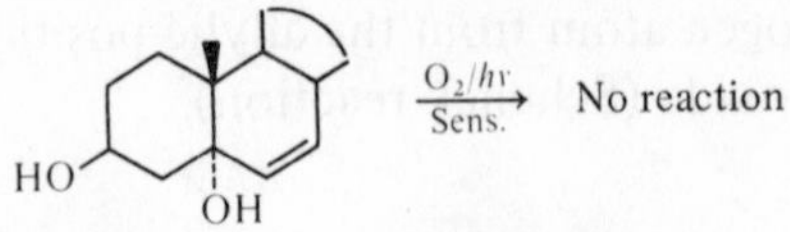

$\xrightarrow[\text{Sens.}]{O_2/h\nu}$ No reaction

It would appear that the best way of explaining the experimentally observed stereospecificity is an attack on the double bond, concerted with the detachment of the allylic *cis*-axial proton, which is perpendicular to the plane of the double bond.

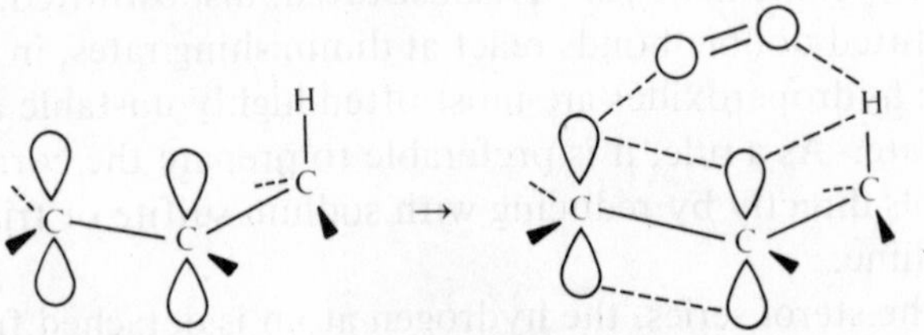

In the sterol series, photosensitized oxidation of an allylic alcohol yields an unstable enol-hydroperoxide, which is rearranged to form an α, β-epoxy ketone:

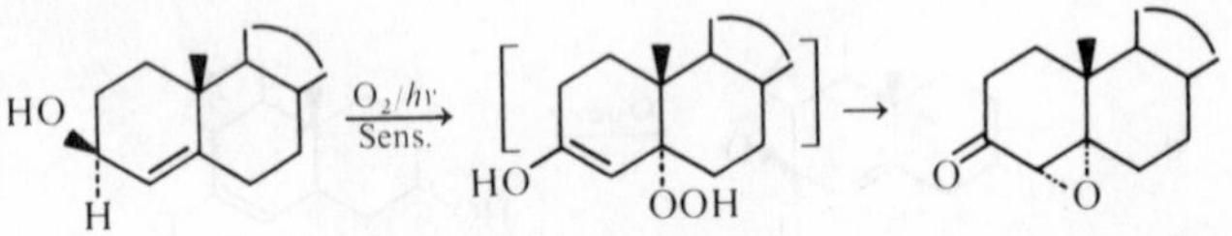

Photosensitized oxidation of β-ionylidene acetic esters yields an exocyclic peroxide, together with a hydroperoxide which is a product of the attack on the tetrasubstituted 5,6 double bond:

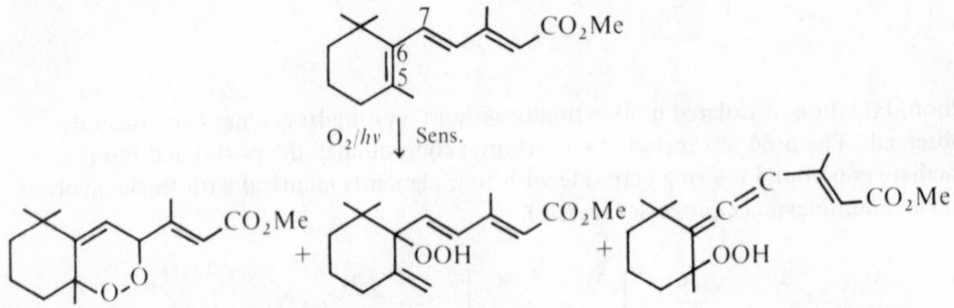

In this case the hydroperoxide formed has an allene structure, and appears to have been formed by the attack on the 5,6 double bond and detachment of the ethylenic hydrogen in C_7. This is the only known example of the formation of an allenic hydroperoxide, but this type of photooxidation can be related to the biosynthesis of certain carotenoids (see next section).

V. BIOLOGICAL IMPLICATIONS OF PHOTOSENSITIZED OXIDATIONS

As was predicted by Dufraisse, in view of the abundance of oxygen and light in nature, photosensitized oxidations are highly important reactions. Reactions which occur in vivo require an adequate sensitizer to take place at all. These conditions are met, inter alia, at the level of chloroplasts or organs with similar functions such as the grana of photosynthetic bacteria. Molecules not absorbing visible light combine with colored pigments such as chlorophyll, phycobiline, phytochrome, and carotenoids at this level.

The part played by carotenoids in sensitized photooxidations has been discovered only recently; it appears that one of the functions of carotenoids is to protect the living cells against a photooxidation sensitized by chlorophyll or by other pigments. Thus, in the absence of carotenoids, the simultaneous action of light, oxygen, and photosensitive pigments results in the destruction of certain bacteria which are not destroyed in the presence of carotenoids. It appears that these molecules act as traps for the singlet oxygen originating from a transfer of the triplet from chlorophyll to the oxygen in the ground state; this is a photosensitized oxidation of carotenoids. The oxidized carotenoids may also return to their initial state: such cycles were studied by Yamamoto, who showed that there are reversible interconversions from violaxanthine to antheraxanthine and to zeaxanthine, the oxidation reactions being photochemical, while the reduction reactions are enzymatic.

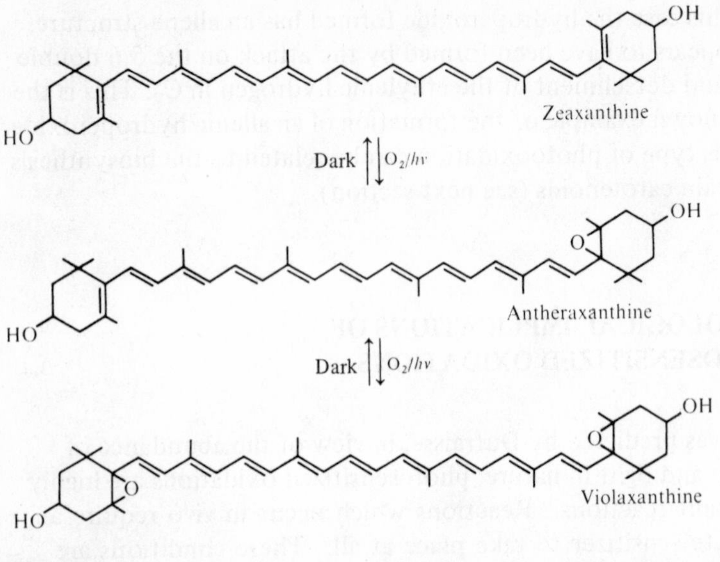

Zeaxanthine

Dark ⇅ $O_2/h\nu$

Antheraxanthine

Dark ⇅ $O_2/h\nu$

Violaxanthine

Recent results showed that the structure of the rearrangement products of photoxides of carotenoid-like polyenes is similar to that of certain natural carotenoids. Examples are rearrangement products of peroxides or hydroperoxides of β-ionylidene acetic ester

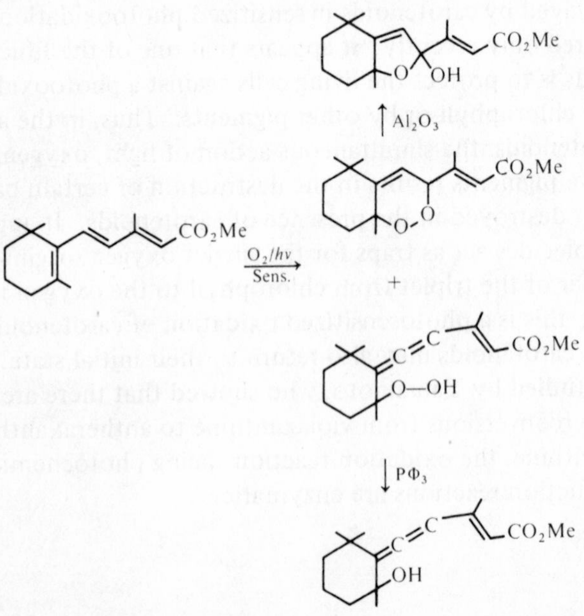

CO_2Me $\xrightarrow[\text{Sens.}]{O_2/h\nu}$

Al_2O_3

$P\Phi_3$

and peroxides of dehydro-β-ionylidene acetic esters

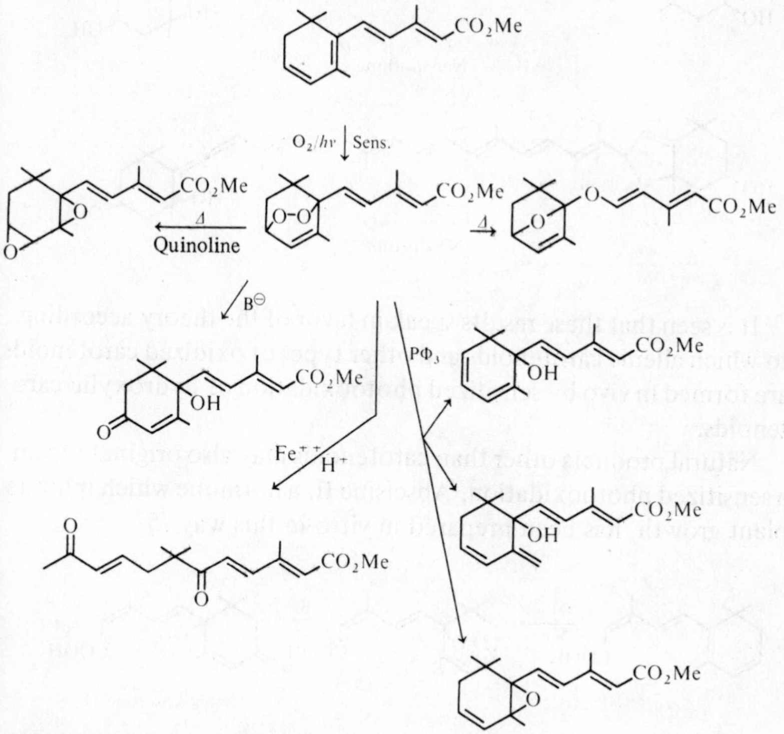

CO_2Me

$O_2/h\nu$ Sens.

CO_2Me Δ Quinoline O CO_2Me O–O Δ O CO_2Me

$B^{\ominus}$ $P\Phi_3$ CO_2Me OH

CO_2Me OH O

Fe^{++} H^+

CO_2Me OH

O CO_2Me O

CO_2Me O

The structure of these compounds is paralleled by the structures of natural carotenoids, viz.:

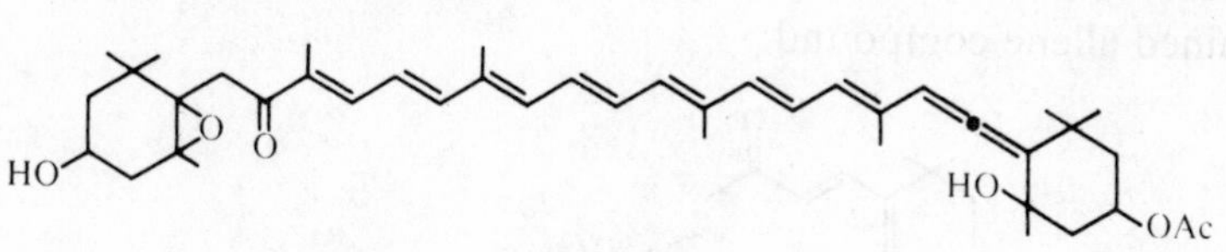

HO O O HO OAc

Fucoxanthine

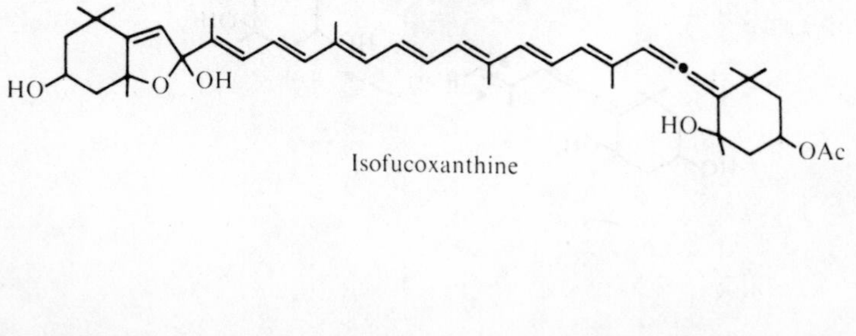

HO O OH HO OAc

Isofucoxanthine

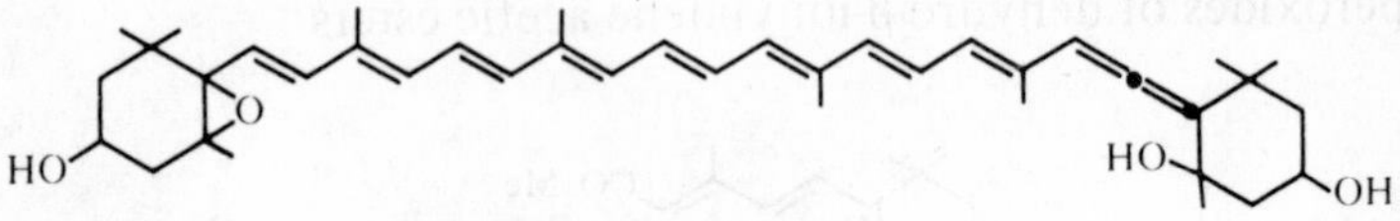

Neoxanthine

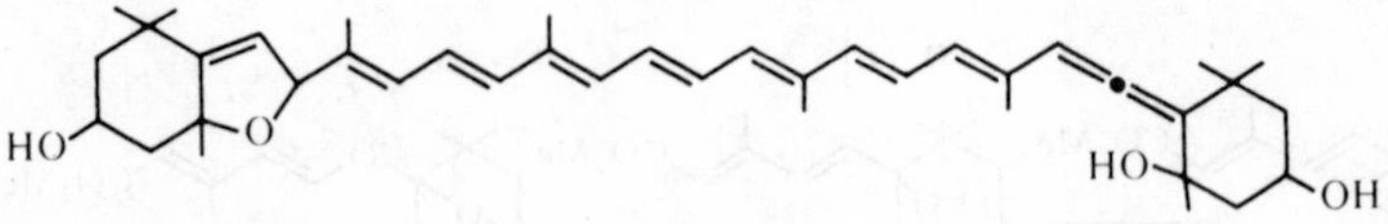

Neochrome

It is seen that these results speak in favor of the theory according to which allenic carotenoids and other types of oxidized carotenoids are formed in vivo by sensitized photooxidation of hydroxylic carotenoids.

Natural products other than carotenoids may also originate from a sensitized photooxidation. Abscisine II, a hormone which inhibits plant growth, has been prepared in vitro in this way /5/:

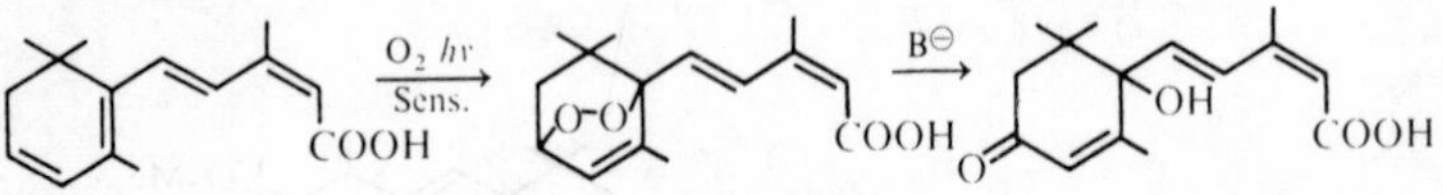

O_2 $h\nu$ Sens. — B$^\ominus$

(±) Abscisine II

Another allenic natural compound (I) has been recently discovered in certain arthropoda. Its synthesis may consist in a photooxidation of 3-hydroxy-β-ionone or in the ozonolysis of a longer-chained allene compound:

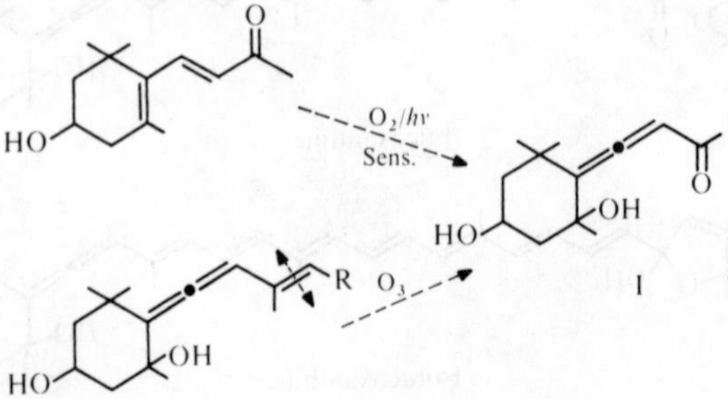

$O_2/h\nu$ Sens. — O_3

I

Two of the products below (II and III), which are responsible for the aroma of black tea, have a structure which is directly related to that of rearranged β-ionone peroxides. They are obtained by degradation of β-ionone peroxide or of higher homologs such as retinyl acetate. The synthesis of theaspirone (III) may have as its starting product the isomer (IV) of β-ionone.

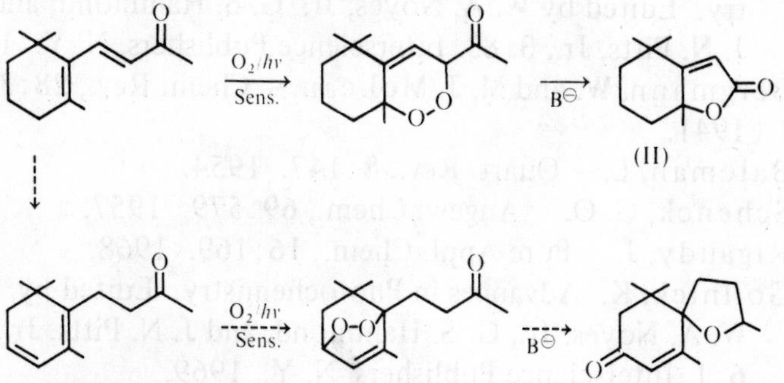

$O_2/h\nu$, Sens. → B⊖ → (II)

(IV) $O_2/h\nu$, Sens. → B⊖ → (III)

Other compounds which must be mentioned include loliolide (V) which is extracted from lilies and other plants, and actinidiolide (VI) and actinidol (VII), which are extracted from *Actinidia polygama:*

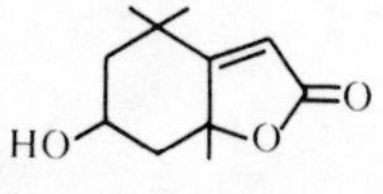

(V) 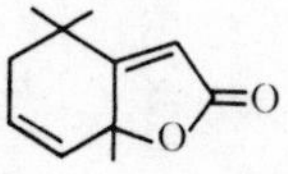 (VI) 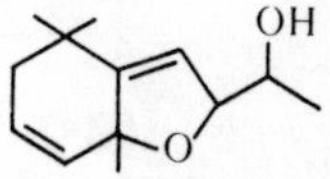 (VII)

Bibliography

1. Foote, C. S. – Acc. Chem. Res., **1**:104. 1968.
2. Gollnick, K. and G. O. Schenck. 1,4-Cycloaddition Reactions. Edited by J. Hamer, p.256. Academic Press, Inc., N. Y. 1967.

3. Griffith, J. S. Oxygen in Animal Organism, p.141. Macmillan, N.Y. 1964.
4. Gollnick, K. and G. O. Schenck. – Pure Appl. Chem., **9**:507. 1964.
5. Mousseron, M., J. P. Dalle, and J. C. Mani. – Bull. Soc. Chim.France, p.1561. 1968; p.232. 1969.
6. Hoare, D. E. and G. S. Pearson. Advances in Photochemistry. Edited by W. A. Noyes, Jr., G. S. Hammond, and J. N. Pitts, Jr., **3**:83. Interscience Publishers, N. Y. 1964.
7. Bergmann, W. and M. J. McLean. – Chem. Rev., **28**:367. 1941.
8. Bateman, L. – Quart. Rev., **8**:147. 1954.
9. Schenck, G. O. – Angew. Chem., **69**:579. 1957.
10. Rigaudy, J. – Pure Appl. Chem., **16**:169. 1968.
11. Gollnick, K. Advances in Photochemistry. Edited by W. A. Noyes, Jr., G. S. Hammond, and J. N. Pitts, Jr., **6**:1. Interscience Publishers, N. Y. 1969.

Chapter 9

PHOTOCHROMISM

A photochromic (phototropic) reaction is a photochemical reaction resulting in a color change of the substrate which can be reversed by interrupting the radiant excitation, this reversion being a thermal process. If a photochromic substance is irradiated with an UV or visible light, a thermodynamically unstable colored substance will be formed, which will be reconverted to the original compound in the dark. Most reactions of this type involve intramolecular cleavage of hydrogen. Photochromic reactions have been observed in liquids and in solids (crystals and glasses).

I. MECHANISM OF PHOTOCHROMIC REACTIONS

The absorption of light by a photochromic substance results in the formation of a "quantum-stable" but "thermodynamically unstable" substance. This is not a mere electronic excitation of a compound, but a true photochemical reaction, yielding different chemical species in its ground state. Phosphorescence and fluorescence phenomena are not in any way connected with photochromism. The compound obtained by a photochromic reaction is most often an isomer of the irradiated compound; since it is thermodynamically unstable in the absence of light, it will be reconverted to the initial compound in the dark:

$$A_0 \xrightarrow{h\nu} A^*$$

$$A^* \longrightarrow B_0$$

$$B_0 \xrightarrow{\Delta} A_0 \,.$$

This is usually written as

$$A_0 \underset{\Delta}{\overset{h\nu}{\rightleftarrows}} B_0 \,.$$

Obviously, such a mechanism does not necessarily mean that the substrate must change color; photochromic reactions are merely a special case of reversible photochemical reactions, and whether or not there will be a color change will depend only on the identity of the substrate.

B_0 can sometimes be also reconverted to A_0 by a photochemical reaction if it absorbs light of a different (mostly longer) wavelength than does A_0. The latter compound can then sometimes be obtained by selective irradiation of B_0:

$$B_0 \xrightarrow{h\nu'} B^*$$

$$B^* \longrightarrow A_0 \,.$$

1. Lifetime of B_0

The unstable product B_0 of a photochromic reaction will be reconverted sooner or later into A_0; the observed lifetimes vary from a few milliseconds to several days. Photochromic reactions lasting for only a few milliseconds have been observed with the aid of flash photolysis. Any practical applications of photochromic reactions will depend entirely on the lifetime of B_0.

2. "Fatigue"

A photochromic substance cannot be utilized indefinitely. The phenomenon of "fatigue" is the main obstacle to an extensive use of these substances. We must distinguish between two types of "fatigue".

A) *Number of reaction cycles*

A large number of photochromic reactions undergone by a substance may eventually modify its properties. However, this type of fatigue is mostly negligible.

B) *Photodegradation*

A photochromic reaction may be accompanied by a secondary irreversible photochemical reaction yielding a stable product. This product may be inert, but may also interfere with the photochromic reaction (chemical reaction, absorption of useful radiation, energy transfer, etc.). This secondary reaction may be due to the fact that the light used in the irradiation is not fully monochromatic and is absorbed by the unstable isomer B_0. Secondary reactions can sometimes be limited in scope by altering the experimental conditions (identity of solvent, presence of oxygen, monochromatic radiation, etc.).

C) *Sensitized photochromic reactions*

A photochromic substance may be excited by the classical kind of photochemical energy transfer.

II. PHOTOCHROMISM IN SOLUTION

The coloration originates from an increase in the degree of conjugation of the photochromic substance during the irradiation, or else is due to the formation of colored free radicals. The degree of conjugation may be increased by several types of reactions:

a) elimination of a substituent;
b) intramolecular hydrogen scission;
c) *cis-trans* isomerism;
d) ring opening.

1. Elimination of a substituent

The product is a quinonoid form of the aromatic nucleus.

A) *Leucocyanides*

Colorless $\underset{\Delta}{\overset{h\nu}{\rightleftarrows}}$ Colored

B) *Camphor derivatives in chloroform solution*

$+ CHCl_3$ $\xrightarrow{-3\ ClH}$

R = Camphor

$O_2/h\nu$

Colored $\underset{\Delta}{\overset{h\nu}{\rightleftarrows}}$ Colorless

2. Intramolecular hydrogen scission

A) *o-Nitro derivatives*

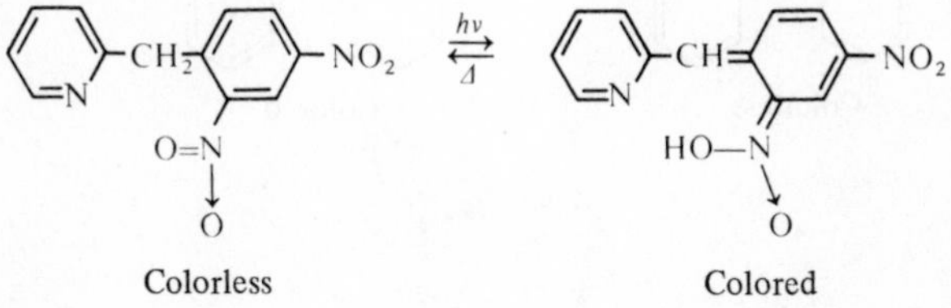

CH_2 NO_2 N O=N O $\underset{\Delta}{\overset{h\nu}{\rightleftarrows}}$ CH= NO_2 N HO—N O

Colorless Colored

B) *Anils*

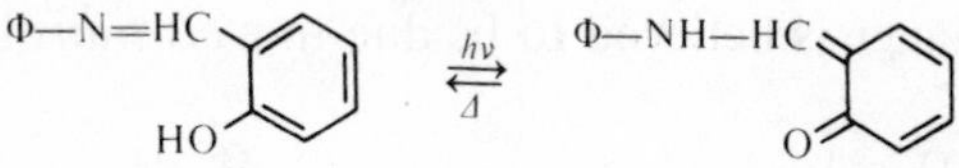

Φ—N=HC HO $\underset{\Delta}{\overset{h\nu}{\rightleftarrows}}$ Φ—NH—HC O

The strongly colored quinonoid form is the result of an intramolecular hydrogen scission. This reaction can only be observed in glasses or in solids (*vide infra*), but also takes place in liquid ethanol solutions; in this case the regeneration of the anil is so fast that the reaction can only be identified by flash photolysis.

C) *Photoenolization of aromatic ketones*

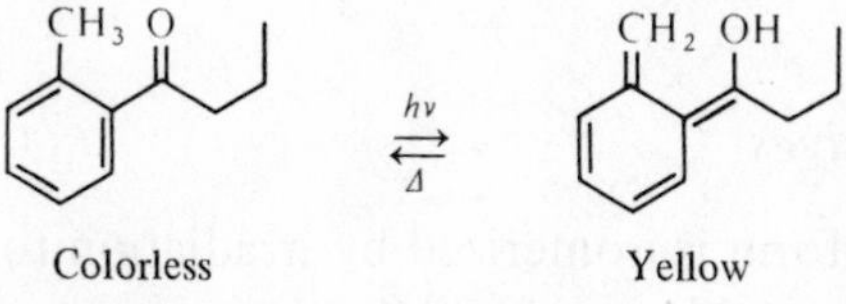

CH_3 O $\underset{\Delta}{\overset{h\nu}{\rightleftarrows}}$ CH_2 OH

Colorless Yellow

o-Methylbutyrophenone reacts differently from the *m*- and *p*-isomers; it does not eliminate ethylene by a type II Norrish reaction (Chapter 7), but gives a reversible photoenolization. At 77°K, the yellow photoenol is stable and can be observed.

This type of photochromism has been observed for other *o*-substituted aromatic ketones, in particular for *o*-benzylbenzophenone:

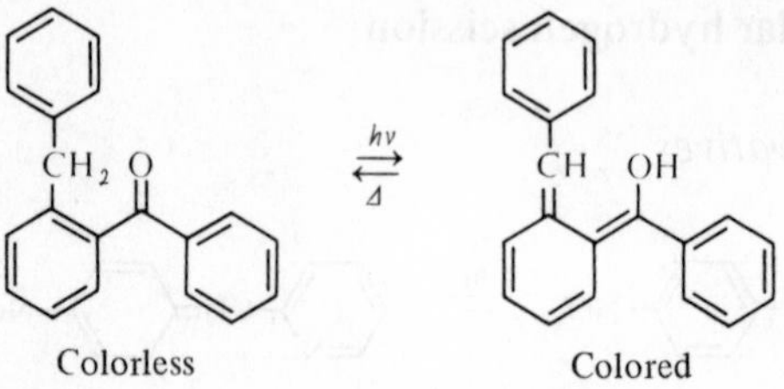

CH$_2$ O $\underset{\Delta}{\overset{h\nu}{\rightleftarrows}}$ CH OH

Colorless   Colored

3. *cis-trans-* Isomerism

A) *Thioindigo dyes*

The color change is believed to be due to a *cis-trans* isomerization:

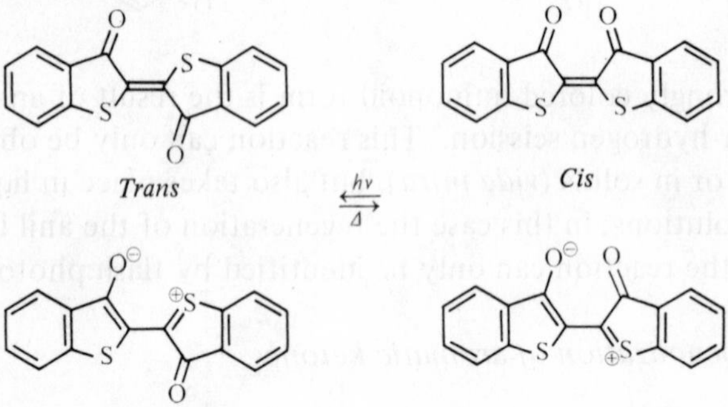

O S S O   O O S S

Trans $\underset{\Delta}{\overset{h\nu}{\rightleftarrows}}$ *Cis*

$O^{\ominus}$ $^{\oplus}S$ S O $O^{\ominus}$ O S $S^{\oplus}$

B) *Aromatic azo dyes*

The stable *anti*-form is isomerized by irradiation to give a *syn*-form, which gives a stable isomer in the dark. In the *o*-hydroxylated derivatives, the *anti*-form is stabilized by a hydrogen bond:

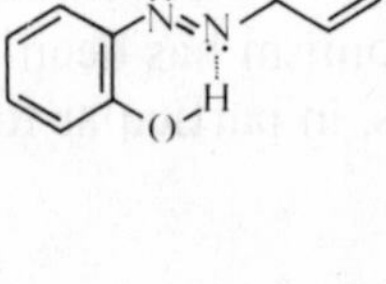

N=N H O

Unsubstituted azobenzene also gives an *anti* $\overset{h\nu}{\rightleftarrows}$ *syn* photochromic reaction.

Crystals of the *syn*-isomer may be stored in the dark, but in solution the conversion of the *syn*- to the *anti*- form is quantitative.

4. Ring opening

A) *Spiropyrans*

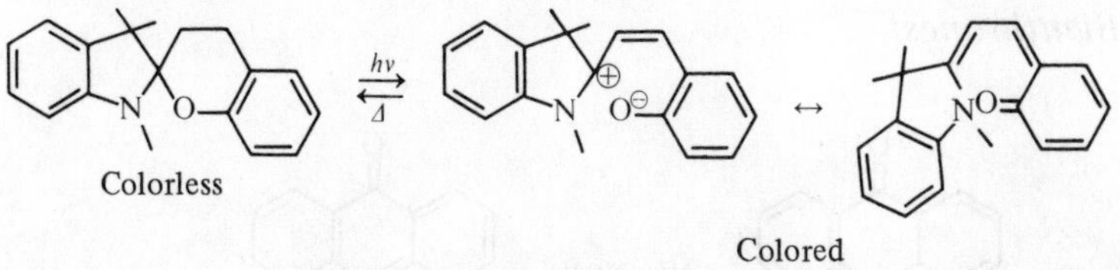

$\overset{h\nu}{\underset{\Delta}{\rightleftarrows}}$

Colorless

Colored

This reaction has been observed for several substituted spiropyrans. The colored form is produced by UV irradiation, while the return to the colorless form is a thermal process, or else is produced by irradiation with visible light.

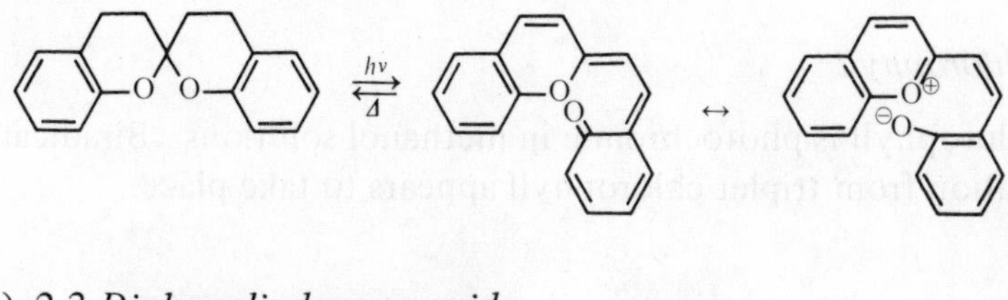

$\overset{h\nu}{\underset{\Delta}{\rightleftarrows}}$

B) 2,3-*Diphenylindenone oxide*

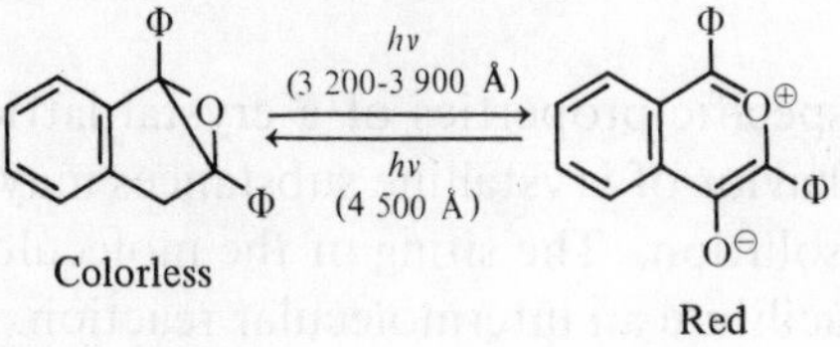

$h\nu$ (3 200-3 900 Å)

$h\nu$ (4 500 Å)

Colorless

Red

5. Free radical formation

A) *Tetrachloroketodehydronaphthalene*

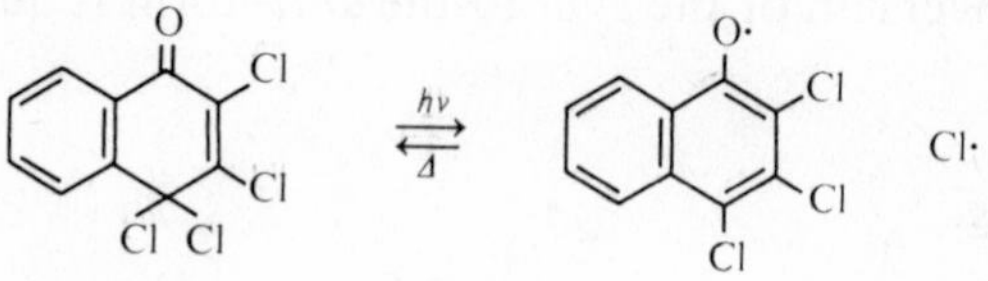

B) *Bianthrones*

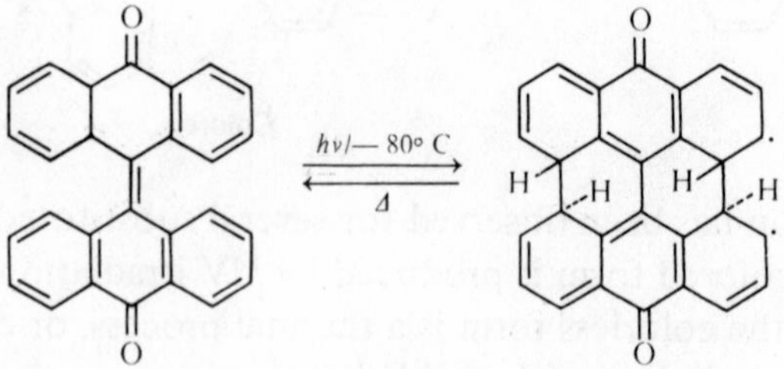

C) *Chlorophyll*

Chlorophyll is photochromic in methanol solutions. Biradical formation from triplet chlorophyll appears to take place.

III. SOLID STATE PHOTOCHROMISM

Owing to the specific properties of a crystal lattice, the photochromic behavior of crystalline substances may differ from their behavior in solution. The siting of the molecules relative to each other may facilitate an intermolecular reaction.

1. Anils

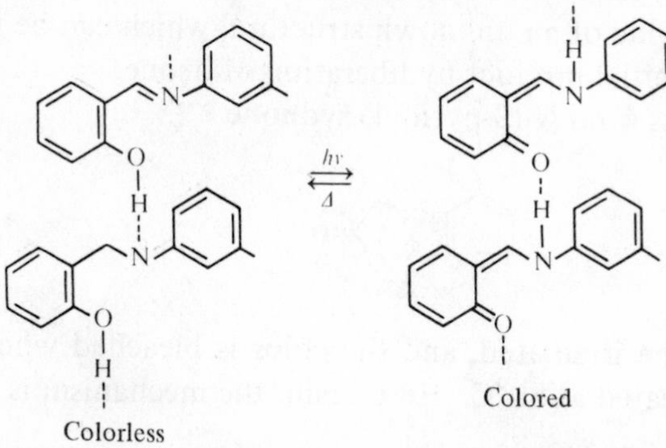

Colorless ⇄ Colored

It is assumed that the molecules in the crystal lattice are situated in parallel planes and that the hydroxyl group of one molecule is close to the nitrogen atom of another molecule. The stable form is colorless, while the *o*-quinonoid form is unstable; the latter form is red or yellow-colored, depending on the identity of the substituents.

2. 2-(2′, 4′-dinitrobenzyl)-pyridine

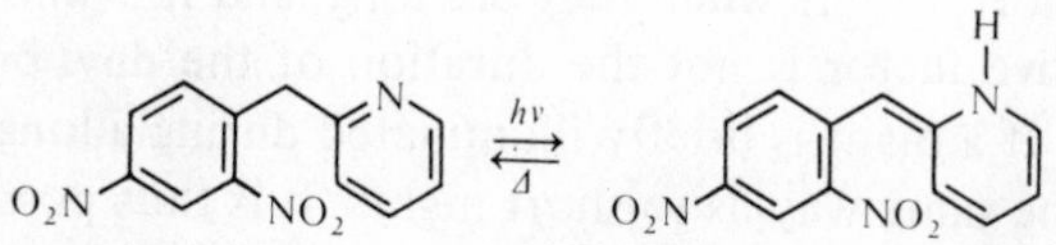

In the crystal lattice the molecules are so situated that the H and N atoms are close to one another, as in anils.

3. Other crystalline photochromic substances

Many hydrazones, osazones and semicarbazones are photochromic in their solid state, but the mechanisms of these reactions are insufficiently understood.

The photochromism of certain substituted stilbenes in the presence of oxygen is probably due to photooxidative formation of a colored oxide of an unknown structure, which can be reconverted to the initial product by liberation of ozone.

The colorless solid N-(3-pyridyl)-sydnone

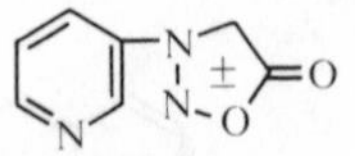

turns blue when irradiated, and the color is bleached when the substance is heated at 80°C. Here again, the mechanism is unclear.

IV. CONTROL OF METABOLISM BY PHOTOCHROMISM

It is known that photochromic reactions take part in the metabolism of certain plants. It is a very familiar fact that under natural conditions certain physiological functions of an organism may display a day-night periodicity; plants and animals are able to recognize the seasons by measuring the duration of days and nights ("biological clock"). Thus, certain plants flower when the days are short, others when they are long, and it is known that the operative factor is not the duration of the day, but that of the night. If a plant is briefly illuminated during a long night, it reacts in the same way as to short nights; it is thus possible to bring about or to prevent the flowering of plants outside their natural periods.

Studies were made on the effect of monochromatic radiations on the flowering of photoperiodic plants. It is interesting to note that a large number of different plants have identical active spectra: red light is the most effective in promoting germination, leaf growth, pigment formation in plant tissues, etc. It would thus seem that all these responses are controled by the same compound, which absorbs in the red. A very careful study of the germination of certain seeds showed that the germination rate increased considerably as a result of irradiation with light of 6600 Å wavelength, but that, on the contrary, irradiation with 7300 Å light inhibited the previously stimulated germination. The inhibiting effect of far red light has been noted for various physiological processes such as flowering, leaf growth, pigment synthesis, etc. Thus, while irradiation with far red light completely inhibits the flowering of a plant, irradiation with red light will stimulate flowering. The system red – far red was demonstrated to be completely reversible for numerous systems: these results indicate that a pigment (phytochrome) undergoes reversible conversion from the form which absorbs at 6600 Å (P 6600) to the form which absorbs at 7300 Å (P 7300):

$$\text{P 6600} \underset{h\nu\,(7\,300\,\text{Å})}{\overset{h\nu\,(6\,600\,\text{Å})}{\rightleftarrows}} \text{P 7300}\,.$$

The P 7300 form is required to initiate the germination process; the stable P 6600 form is the only one which exists in the dark. Irradiation at 6600 Å yields the P 7300 form, which initiates germination; a subsequent irradiation with 7300 Å results in the regeneration of the P 6600 form, thus restoring the dormancy. Physicochemical studies of phytochromes showed that the two forms have similar extinction coefficients at maximum absorption, but very different quantum yields: the conversion of the P 6600 form to the P 7300 form by 6600 Å radiation is at least four times more efficient than the reverse conversion produced by 7300 Å light. This difference in the quantum yields is a very important factor in the photoperiodicity of plants. It has been proved that P 7300 is spontaneously converted to P 6600 in the dark and that the reaction is in fact photochromic:

$$\text{P 6600} \underset{\Delta}{\overset{h\nu\,(6\,600\,\text{Å})}{\rightleftarrows}} \text{P 7300}\,.$$

This slow conversion in the dark is thought to be the way of measuring the time. When a plant has been irradiated at 6600 Å, its P 6600 pigment is converted to P 7300; in the dark, P 7300 will be reconverted to P 6600 at a rate which varies in individual plants. The processes which occur only during long nights must be assumed to be due to the presence of the P 6600 phytochrome alone; if the night is briefly interrupted by irradiation with 6600 Å light, P 6600 is reconverted to P 7300, after which the transformations taking place in the dark are resumed. The net effect is that a long night has been replaced by two short nights; the effect of this interruption can be suppressed by subsequent irradiation at 7300 Å, since this radiation produces the P 6600 pigment much faster than the thermal reaction in the dark.

The fact that sunlight acts first and foremost as a red light must be due to the higher quantum yield of the 6600 Å reaction and to the fact that plants store phytochrome in the dark in the P 6600 form. The effect of the far red sunlight fraction can be observed if the red light is filtered by the chlorophyll of the leaves, which transmits more far red than red light: thus, in a thick forest, seeds which receive only far red light remain dormant; they only germinate when exposed to direct sunlight.

The biological role of the two phytochrome forms is as yet unclear. Many hypotheses concerning the participation of phytochrome in the germination process have been advanced. Germination is probably due to the breaking of the seed hull; this may be effected by the pigment P 7300 (produced by the action of red light on P 6600) which activates the hull breaking enzyme. The same result is produced by gibberellin, an α-amylase-activating hormone which degrades the polysaccharides in the hull. According to another hypothesis, the pigment P7300 has a sensitizing effect on the enzyme which destroys the growth inhibitor in the hull.

Red light could possibly play the following part in the color changes taking place in the leaves. In summer, the chlorophyll-rich leaves are green; the red light is filtered, and the phytochrome remains in the P 6600 form. When the chlorophyll loses its color, the leaves become yellow owing to the presence of carotenoid pigments. The red light converts the phytochrome into P 7300 which in turn accelerates the synthesis of anthocyanins which are responsible for the red color of the leaves. The synthesis of these anthocyanins is closely connected with the plant metabolism: the glycogen is utilized as glucose-1-phosphate. In the absence of phosphates, the excess glucose is used in the synthesis of anthocyanin. This dearth of phosphates may be due to leanness of the soil (red leaves in phosphate-poor soils). In autumn, the decreased phosphate concentration in the plant nutrient elements and the decoloration of the chlorophyll brings about the synthesis of anthocyanins.

Bibliography

1. Dessauer, R. and J. P. Paris. Advances in Photochemistry. Edited by W. A. Noyes, Jr., G. S. Hammond, and J. N. Pitts, Jr., 1:275–322. Interscience Publishers, N.Y. 1963.
2. Fischer, E. – Fortsch. Chem. Forsch. (Photochemie), 7:605. 1967.
3. Ullman, E. F. et al. – J. Am. Chem. Soc., 87:5417, 5424. 1965.
4. Douzou, P. and C. Wippler. – J. Chim. Phys., **60**: 1409. 1963.
5. Exelby, R. and R. Grinter. – Chem. Rev., **65**:247. 1965.
6. Luck, W. and H. Sand. – Angew. Chem., **76**: 463. 1964.
7. Seliger, H. H. and W. D. McElroy. – Light: Physical and Biological Action, p.248. Academic Press, N.Y. 1965.

Chapter 10

CHEMILUMINESCENCE AND BIOLUMINESCENCE

We have so far dealt with excited states produced by an initial absorption of a photon by a molecule, either by direct excitation or by energy transfer. However, the energy which is required for excitation may also originate from other sources; thus if the intermediate state of a chemical or a biochemical reaction has a sufficiently high energy level, the reaction product may be obtained in an excited state. If the excited molecule emits a photon in returning to the ground state, we speak of a chemiluminescent reaction. Such reactions usually consist in the emission of fluorescence; if the reaction is biochemical, we speak of bioluminescence. The excited molecule may sometimes transmit its energy to another molecule which then emits luminescence; this is sensitized chemiluminescence. The chemiluminescence quantum yields (number of photons emitted per molecule destroyed are very low ($10^{-5} - 10^{-15}$), while those of bioluminescence may be as high as 1.

Many chemiluminescent reactions are known, but their mechanisms and their emission spectra have not yet been studied to a significant extent.

I. CHEMILUMINESCENCE

1. Reactions involving peroxides

It is important to note that most organic chemiluminescent reactions involve molecular oxygen and peroxides. The oxygen which is liberated as a result of decomposition of peroxides may be assumed to be excited (Chapter 8); in the decomposition of hydrogen peroxide, the red luminescence (6334 and 7032 Å) originates from an excited dimer 2 $^1\Delta_g$. The different electronic states of oxygen and their dimeric combinations furnish a scale of energy levels ranging from 23 to 83 kcal·mole^{-1}. According to a recent theory of chemiluminescence, an energy transfer takes place between the liberated excited oxygen and the acceptor molecule, which emits the luminescence. Several facts can be quoted in support of this theory:

a) the chemiluminescence spectrum is usually similar to the emission spectrum of the original precursor molecule;

b) the chemiluminescence of the acceptor molecule varies regularly with the peroxide concentration;

c) the very low quantum yield of the chemiluminescence indicates that it is only a minor effect as compared to the main oxidation reaction.

2. Other types of chemiluminescence

Other chemiluminescent reactions do not involve peroxides; these are very vigorous free radical or ionic reactions.

A) *Formation of single bonds between two radicals*

$$R\cdot + R'\cdot \rightarrow R{-}R' + h\nu \, .$$

Example: $NO + O \rightarrow NO_2 + h\nu$

B) *Formation of double bonds between two biradicals*

$$\cdot R\cdot + \cdot R'\cdot \rightarrow R{=}R' + h\nu \, .$$

This case includes carbene duplication or oxidation:

$$2CH_2{:} \rightarrow CH_2{=}CH_2 + h\nu \, .$$

During the oxidation of diphenylcarbene to benzophenone, the observed luminescence is the phosphorescence emitted by benzophenone:

$$(C_6H_5)_2\ddot{C} \xrightarrow{O_2} ((C_6H_5)_2C{=}O)^* \longrightarrow h\nu$$

C) *Electron capture*

$$R^+ + e^- \rightarrow R + h\nu\,.$$

Photoisomerization of certain organic molecules, such as aromatic amines, in solid medium at a low temperature, makes it possible to stabilize the ionic species formed. When the temperature is raised, chemiluminescence is observed owing to the capture of the electron by the cation:

$$R \xrightarrow{h\nu} R^+ + e^-$$
$$R^+ + e^- \xrightarrow{\Delta} R + h\nu'\,.$$

The observed luminescence is usually the phosphorescence emitted by R; sometimes both fluorescence and phosphorescence are observed. The phosphorescence seems to originate from a direct excitation to the triplet state rather than from an intersystem crossing starting from the excited singlet state.

3. Examples of chemiluminescence

Luminol. When hydrogen peroxide is made to act on luminol in a basic medium, chemiluminescence is observed; this luminescence is identical with the fluorescence of aminophthalate ion:

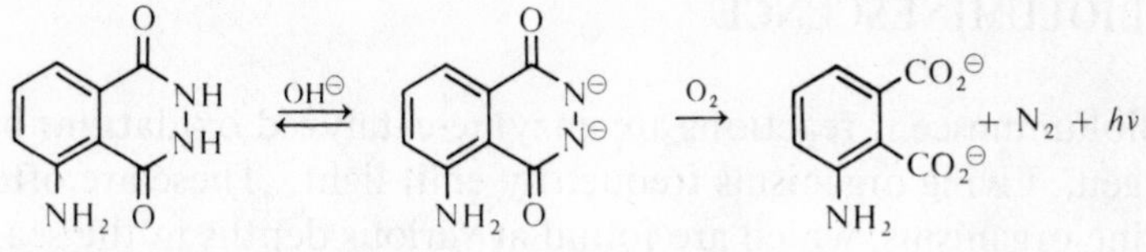

Luminol

Lophines. The chemiluminescence of lophine (2,4,5-triphenylimidazole) is observed when oxygen is allowed to act on alkaline solutions of lophine. The reaction very probably proceeds via a hydroperoxide intermediate. The observed emission is the fluorescence of amidine ion:

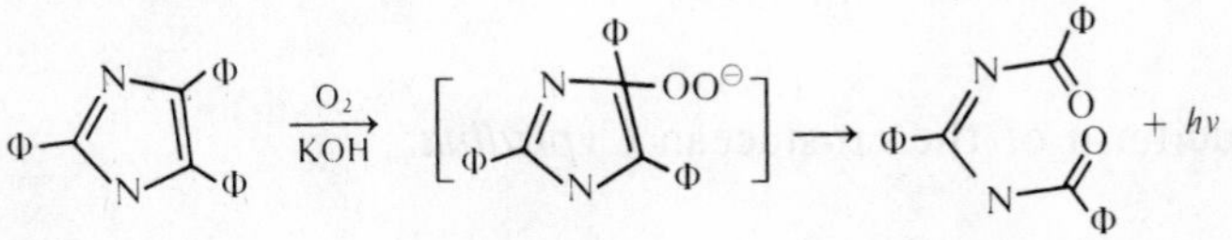

Indole compounds. Chemiluminescence of indole compounds has been intensely studied during the past few years, mainly owing to the establishment of the structure of the various luciferins; these compounds contain an indole nucleus and are bioluminescent (*vide infra*).

Chemiluminescent reactions of indole also proceed via a peroxide intermediate:

$\rightarrow h\nu$ + products

$\rightarrow h\nu$ + products

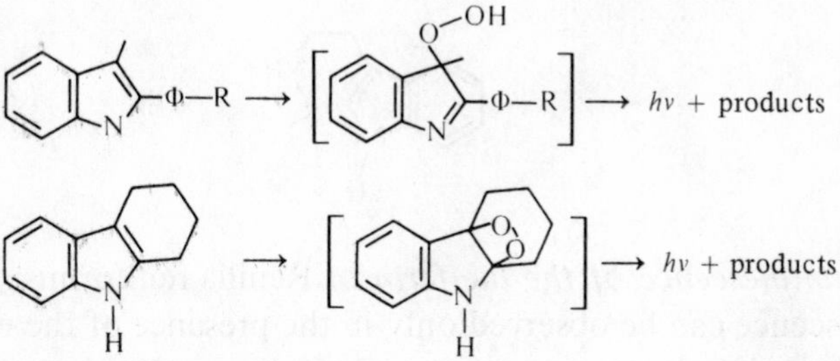

Since all the reaction products are fluorescent, it is difficult to identify the emitting species.

II. BIOLUMINESCENCE

Bioluminescent reactions are enzyme-catalyzed oxidations by oxygen. Living organisms frequently emit light. These are often marine organisms, which are found at various depths in the sea.

Three species must be present for the light to be emitted: a) an oxidation-catalyzing enzyme (luciferase); b) the oxidized substrate (luciferin), and c) molecular oxygen.

Certain luciferins have been isolated and studied. These include the luciferin of firefly:

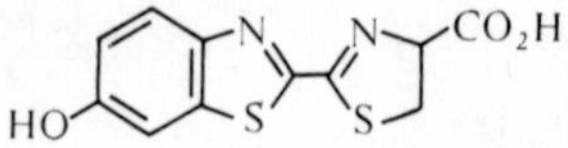

the luciferin of the crustacean *Cypridina:*

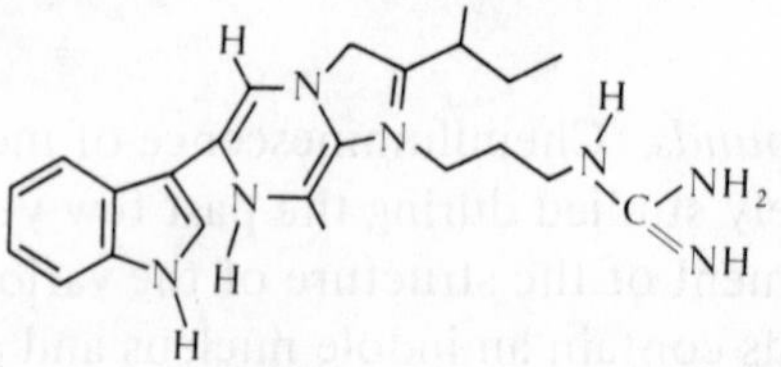

and the luciferin of *Renilla reniformis,* the bioluminescence of which will be discussed below, and which is a derivative of tryptamine:

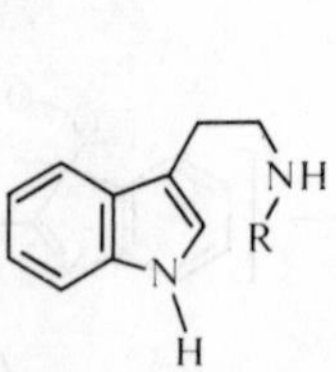

Bioluminescence of the luciferin of Renilla reniformis. The luminescence can be observed only in the presence of the enzyme (*Renilla* luciferase), the coenzyme (3′,5′-diphosphoadenosine), Ca^{++} ions, luciferin, and oxygen.

$$\text{Luciferin + coenzyme} \xrightarrow[\text{Ca}^{++}]{\text{Luciferase}} \text{Activated luciferin.}$$

It is important to note that this form of activated luciferin can also be obtained by heating in an acid medium, without the enzyme, the coenzyme, and Ca^{++} ions. The chemiluminescence which is thus obtained in the presence of oxygen is identical to the bioluminescence, showing that the same activated form of luciferine is involved. Such a form might be the product of hydrolysis of a group (X), as follows:

$$\text{Luciferyl-X} \rightarrow \text{Activated luciferin + X.}$$

It was experimentally demonstrated that the X residue is the sulfate group bound to the R group of tryptamine in *Renilla* luciferin.

The mechanism of the bioluminescence may thus be schematically represented as follows:

$$\text{Luciferyl sulfate + coenzyme} \xrightarrow[\text{Ca}^{++}]{\text{Enzyme}} \text{Luciferin + PAPS}$$

where PAPS is phosphoadenosine phosphosulfate.

$$\text{Luciferin} + O_2 \xrightarrow{\text{Enzyme}} h\nu + \text{products.}$$

If it exists at all, the enzyme involved in the second stage is probably different from that of the first stage.

Bibliography

1. Johnson, F. H. and Y. Haneda, Editors. Bioluminescence in Progress. Princeton University Press, Princeton. 1966.
2. Symposium on Chemiluminescence. – Photochemistry and Photobiology, **4**: 957. 1965.

3. McCapra, F. – Quarterly Reviews, 20 : 485–510. 1966.
4. Bowen, E. J. – Pure Appl. Chem., 9 : 473. 1964.
5. Goto, T. and Y. Kishi. – Angew. Chem. (Int. ed.), 7 : 407. 1968.

Chapter 11

THE VISUAL PROCESS

I. INTRODUCTION

The problem of the visual receptor is simply this: how is optical information converted to electric information? Two answers have so far been offered: G. Wald (Nobel laureate) assumes that the main step involved is photoisomerization of the visual pigment; B. Rosenberg assumes that the process is based on photoconduction of polyene compounds. The latter theory provides a physical basis for Svaetichin's chromatic "S" potentials (electrophysiological responses to colors): variations in the retinal potential – hyperpolarization or depolarization – persist as long as the retina is irradiated by the incident light, and depend on the particular color.

We shall first discuss briefly the structure of the visual receptor.

II. THE VISUAL RECEPTOR

The external segments of the retinal cones and rods contain the visual pigment and comprise 200–1000 disks which are arranged in a highly ordered lamellar structure. These disks, which are 100–200 Å thick, contain almost all the proteins and the lipids of the segment. The main protein component is the visual pigment: 40% rhodopsin in the rods, which also contain

48% lipids. The chromophore (retinal) lies on the surface of the protein (Figure 11.1).

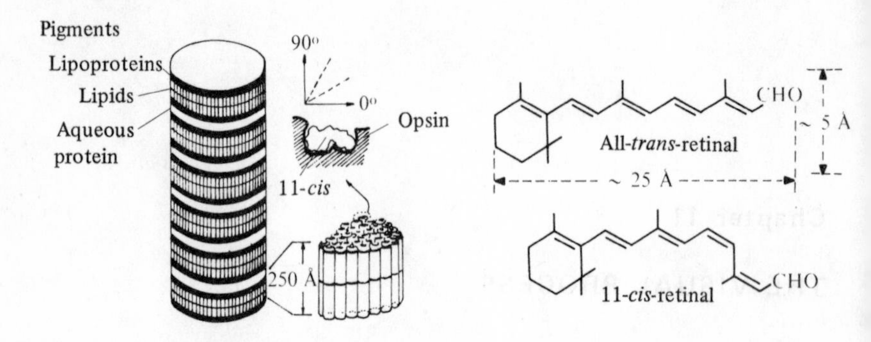

FIGURE 11.1

Three visual pigments have been isolated from the human eye: rhodopsin in the rods, chlorolabe and erythrolabe in the cones. These pigments are Schiff bases, the constituents of which are 11-*cis*-retinal and a protein of molecular weight of 40,000–60,000 (opsin in rods, photopsin in cones). In certain fish species, other pigments – porphyropsin and cyanopsin – have dehydroretinal as chromophore.

III. THEORY OF DECOLORATION OF THE VISUAL PIGMENT (G. Wald)

The decoloration of irradiated rhodopsin, which is observed both in vitro and in vivo, has been studied by several scientists, and primarily by Wald and Hubbard. The retinal which is bound to the protein in the pigment has the sterically hindered configuration 11-*cis*. According to Wald, the only effect which is produced by light is a photoisomerization to an all-*trans*-isomer; the pigment is hydrolyzed by a sequence of thermal or enzymatic processes, with liberation of the all-*trans*-retinal and consequent decoloration. This retinal is then isomerized to 11-*cis*-retinal by retinene isomerase, after which it recombines with the protein to give the pigment.

TABLE 11.1

Reaction	Product	Color
	11-*cis*-Rhodopsin	red
Stereoisomerization of retinal (reversible)	$h\nu$ ↓↑	
	Lumirhodopsin (all-*trans*)	purple-red
Thermal rearrangement of opsin	↓	
	Metarhodopsin	orange-red
Hydrolysis of the chromophore (decoloration)	↓	
	Retinal (all-*trans*) + opsin	yellow

Thus, according to Wald, the decoloration is a direct result of the absorption of light by rhodopsin; retinene isomerase then acts on the all-*trans*-retinal yielding 11-*cis*-retinal, which then recombines with opsin, thus closing the cycle.

The isomerization could then be converted to electric pulses in various manners. The isomerization may produce a continuous solution in the membrane of a disk and depolarize it. Alternatively, opsin may be an enzyme which secretes an active substance, while rhodopsin is inactive, because a part of the active site is covered by retinal or because the protein is deformed by retinal.

Wald's theory met with numerous objections, which showed that other approaches to the problem had to be investigated. The objections are of two kinds.

Firstly, it was found that it is possible to pass from rhodopsin to prelumirhodopsin at −195°C, and from rhodopsin to lumirhodopsin at -65°C, which means that there are no significant differences in structure between these compounds.

Secondly, while the half-regeneration time of opsin is 90 sec, it was shown by Brown and Murakami and later by Smith and Brown that if a retina is flash-irradiated, intracellular currents are observed within a few microseconds, their action spectrum being identical to that of rhodopsin. Such rapid response is incompatible with the transfer and amplification mechanisms proposed by Wald.

The vision of cephalopods does not involve the decoloration of the pigment, showing that the pigment does not play an essential part.

IV. PHOTOCONDUCTION THEORY (B. Rosenberg)

In view of the results just discussed, it was suggested by Rosenberg that the visual mechanism involves a direct conversion of optical information, viz., generation of charge carrier by light. The current, which is produced by the movement of the charge carriers, can modify at a distance the potential of certain membranes. If this is so, the photoisomerization of the visual pigment would be a mere adaptation process, the purpose of which is to enable the eye to respond to a wider range of intensities.

The structure of the visual organs resembles that of other organs such as chloroplasts, muscle fibers, or membranes of nerve myelin. All these organs have a property in common, viz., they form structural assemblies which are ordered on the molecular level. The existence of such structures immediately poses the question of the relationship between the structure and the function of an organ. According to Rosenberg, the structure of the visual organs plays an essential part in their functioning; accordingly, in a study of the process of conversion from optical to electric information, the models used must not only consist of molecules similar to those present in the visual receptor, but must also have a general structure resembling that of the receptor.

Such properties, which result from the arrangement of the molecules, have been detected in a photosynthesis unit, in which the energy captured by one out of 250 chlorophyll molecules is transmitted to a single collector center. A number of facts about the visual receptor show that the interaction between the individual molecules is sufficiently strong for us to be able to treat the termini of cones and rods as "pseudo-solids."

All carotenoid pigments which are Schiff bases of retinal are photoconducting. Rods in vitro are photoconducting in the absence of moisture, as are rods in their normal physiological condition.

1. Experimental study of photoconduction of β-carotene

In these experiments the pigment was compressed between two electrodes transparent to the incident light and connected to a battery. The current produced by the irradiation was measured with the aid of an electrometer. Two types of electric current were detected:

a) Photoconduction current: the amplitude of the current during irradiation is significantly larger than the amplitude of the current in the dark. The photoconduction current is directly proportional to the intensity of the incident light and to the applied voltage (between 20mV and 500V).

b) Photovoltaic current: this current is observed during irradiation when no potential difference is applied to the electrodes. In the case of β-carotene, the photovoltage is about 200mV. The photovoltaic effect is essentially produced by a nonhomogeneous absorption of the light in the bulk of the cell, with consequent nonhomogeneous distribution of the charge carriers. The diffusion of these charge carriers moving across free charge concentration gradients results in the appearance of a photovoltaic current. The stronger the light absorption, the stronger the photovoltaic effect.

These photocurrents are propagated by positive "holes"; they move from the positive to the negative electrode in the photoconduction effect, and in the direction of the incident light in the photovoltaic effect. Their action spectra (i.e., plots of current intensity as a function of the wavelength of the irradiation) have been studied.

The action spectra of the photoconduction current (irradiation through the positive electrode) and of the photovoltaic current are coincident with the absorption spectrum of the pigment.

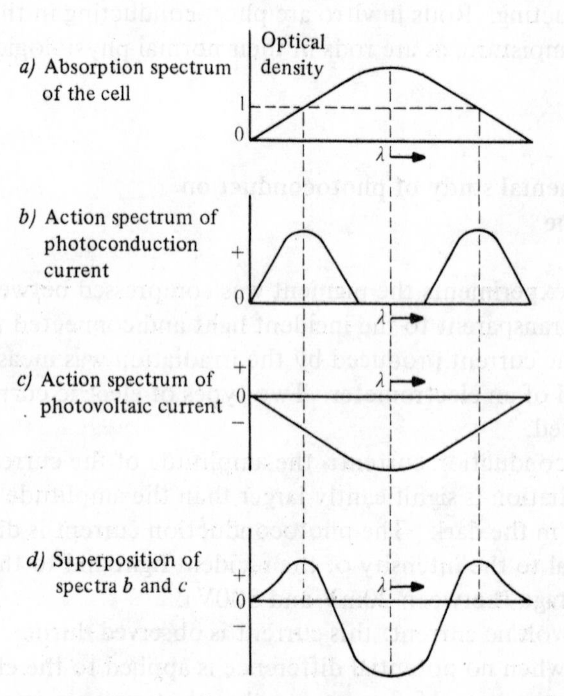

FIGURE 11.2. Schematic action spectrum of a cell of β-carotene irradiated through the negative electrode.

The action spectrum of the photoconduction current (irradiation through the negative electrode) has a peak in the red and is not much affected by wavelengths which are strongly absorbed by β-carotene;

these wavelengths do not reach the positive electrode, while the light which is absorbed near this electrode is the only effective light. A peak of the action spectrum will correspond to wavelengths at which the absorbance of the cell is 0.1 – 0.3 (Figure 11.2).

We have here the effect of self-filtration by a β-carotene cell: when the irradiation is conducted through the positive electrode with a light which has filtered through a solution of β-carotene, we observe a shift of the maximum of the spectrum toward longer wavelengths (i.e., wavelengths which are not significantly absorbed by the solution) as the β-carotene concentration increases.

2. Competition between photoisomerization and photoconduction

If the nature of the irradiated isomer of β-carotene in fact influences the photocurrent, a temperature effect should be observed. This has been verified: the probability of charge generation in the sterically hindered *cis*-isomer is 10^3 times higher than in the non-hindered *cis*-isomer and 10^6 times higher than in an all-*trans*-isomer.

Let us examine the potential energy curves of the different electronic states of a molecule as a function of the rotation angle about the conjugated double bond (Figure 11.3).

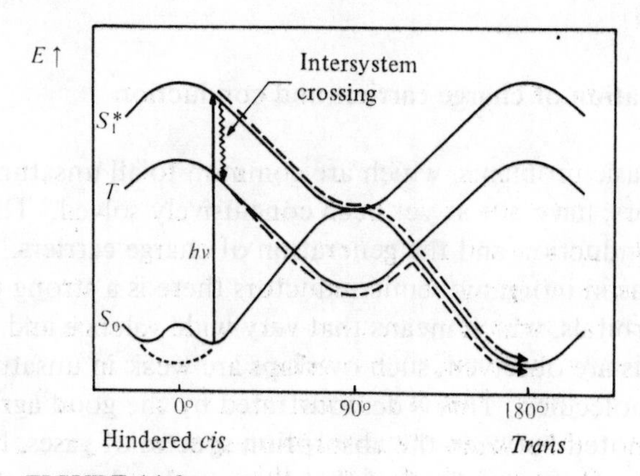

FIGURE 11.3

The thermal isomerization may proceed entirely from the ground state or else from the triplet state if the potential energy curves of the two states intersect. No kinetic data have been reported for retinal. The photoisomerization may proceed through at least two different states: the first excited singlet state and the triplet state.

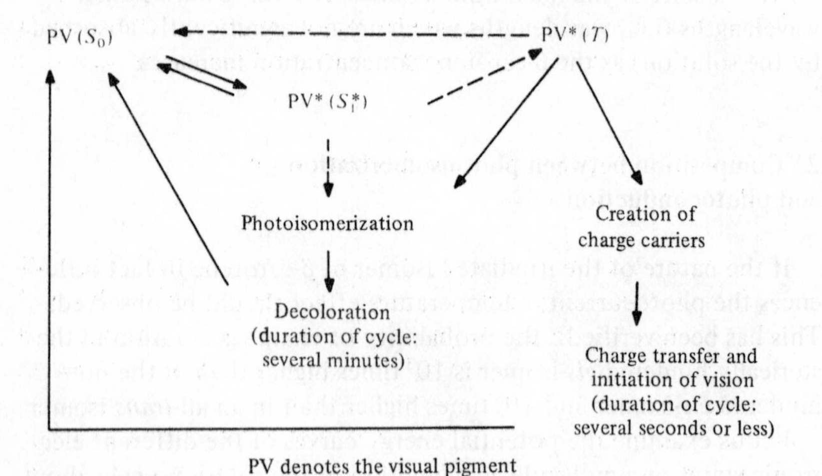

FIGURE 11.4

3. Generation of charge carriers and conduction

Two basic problems, which are common to all unsaturated photoconductors, have not as yet been conclusively solved. They are charge conduction and the generation of charge carriers.

Whereas in inorganic semiconductors there is a strong overlap of atomic orbitals, which means that very wide valence and conduction bands are observed, such overlaps are weak in unsaturated organic molecules. This is demonstrated by the good agreement which is noted between the absorption spectra of gases, liquids and solids. However, the fact that these molecules are semiconducting in the crystalline state means that there are narrow conduction bands produced by a partial overlap of the orbitals. Nevertheless, the possibility of charge carriers jumping from one molecule to another cannot be altogether dismissed.

As in the case of inorganic semiconductors, such charge carriers can be electrons or positive holes; they cannot be excitons (triplet or singlet), contrary to the views of Terenin. Such excitons, which are formed through light absorption by the crystals, can move very rapidly through the crystal lattice, but this does not involve charge transfer. We may imagine two ways in which such charge carriers could be generated:

a) By an external process

$$S_0 + h\nu \rightarrow S^*$$

$$S^* + M \text{ (surface)} \rightarrow C^+ + C^- .$$

b) By an internal process:
indirectly

$$S_0 + h\nu \rightarrow S^*$$

$$S^* + S^* \rightarrow C^+ + C^-$$

or directly

$$S_0 + h\nu \rightarrow C^+ + C^-$$

where S_0 is the molecule in its ground state, S^* is the excited state, M is a molecule not forming part of the lattice, and C^+ and C^- are charge carriers. These two modes of generation are difficult to differentiate, but a few relevant differences may be noted:

a) Variation with the intensity of the incident light.

For a given excited state the steady-state exciton concentration is

$$[S^*] = \epsilon I \tau,$$

where ϵ is the absorption coefficient, I is the intensity of the incident light, and τ is the lifetime of the excited state. An examination of the equations given above will show that the generation of charge carriers will be a first order reaction for an external process, a second order reaction for an indirect internal process and a first order reaction for a direct internal process.

b) Quenching.

Since external generation of the charge carriers involves a transfer of an exciton to an impurity, any process which interferes with the transfer will interrupt the generation of the carriers; the same applies to two-photon processes. On the contrary, one-photon processes will not be affected.

c) Effect of absorption coefficient.

External carrier generation is favored by a high absorption coefficient; an internal one-photon process, on the contrary, which competes with exciton generation, is suppressed by a high absorption coefficient.

Results obtained for the photoconduction of anthracene show that the photoconduction is external at wavelengths above 2800 Å and monophotonic internal at shorter wavelengths. Two-photon processes do not occur except with very powerful sources such as lasers.

It seems that the photoconduction of β-carotene in the visible is due to the injection of holes by the positive electrode. The charge carriers are generated by migration of a nonionized exciton toward the surface of the positive electrode, where it is destroyed; this results in the injection of an electron into the positive electrode or, which is the same thing, in the injection of a hole into the semiconductor. The possible hole injections are a function of the nature of the electrode and of the exciton energy. This hypothesis is supported by the following facts:

a) The positive holes are the main carriers; this means that the generation process is external, or we would have $i^+ \simeq i^-$, since the mobilities of the two types of carriers are similar.

b) The action spectrum of (thin film) photoconduction is identical with the absorption spectrum of β-carotene, which shows that excitons are indeed the kinetic intermediates.

Such charge carrier generation processes have been observed for anthracene and tetracene. β-Carotene represents a special case, since its rectification coefficient is exceptionally high.

4. Amplification by photoconduction

One photon is sufficient to activate a rod and thus to displace 10^5 charge carriers. This amplification cannot be explained by the enzymatic effect of opsin, but photoconduction has an intrinsic gain effect; a hypothetical visual receptor, with a segment length of $20\,\mu$, a voltage of 100 mV (the normal membrane voltage), lifetime of charge carrier 10^{-1} sec (actual range between 10^{-3} sec and 10 sec) and mobility of charge carrier $1\,\text{cm}^2/\text{volt}\cdot\text{sec}$ would have a gain factor of 10^4.

5. Color response

A) *β-Carotene cell*

A β-carotene cell gives a specific response to the absorbed wavelength. There are three experimental conditions which are necessary and sufficient for the cell to detect the color:

a) the two photoelectric processes (photoconduction and the photovoltaic effect), which have different rate constants, must take place simultaneously in a single cell;

b) these two processes may result in currents moving in opposite directions;

c) two different action spectra should correspond to these two opposed currents.

Figure 11.5 is a schematic representation of the different responses of a β-carotene cell which has been irradiated through the positive electrode (a), through the negative electrode (c), and at zero potential difference between the electrodes (b).

Experiment a : the only effective light is blue-green; this light is strongly absorbed by the cell (photoconduction and voltaic current).

Experiment b : the only effective light is blue-green (photovoltaic current alone).

Experiment c : photoconduction and photovoltaic currents have opposite directions. The blue-green light produces the photovoltaic

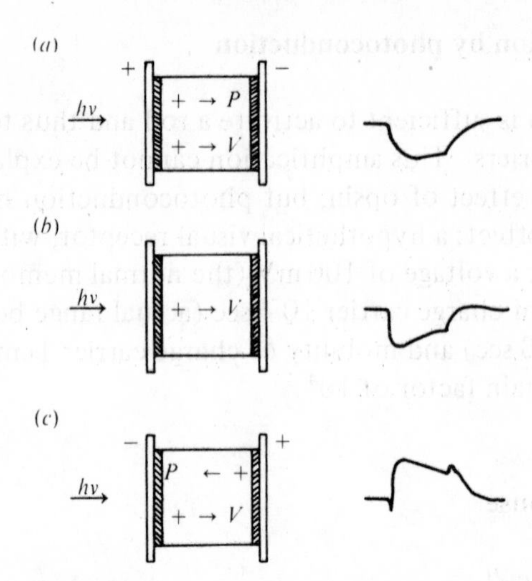

FIGURE 11.5. P – photoconduction current
V – photovoltaic current

current, while the red light (which is only weakly absorbed by the cell) is responsible for the photoconduction. The resulting current will be negative if the photovoltaic effect predominates, and positive if photoconduction predominates. If the two currents have equal amplitudes, the resulting current will be zero and the only effects observed will be the signals indicating the beginning and the end of the irradiation; these signals are due to the photovoltaic effect, since the initial and the final slopes of its response curve are very steep. Studies carried out using monochromatic light yielded signals which were typical of individual colors. The wavelength which produces zero current will depend on the voltage.

For the human eye to perceive the full color scale, one type of cells is insufficient; three types of cells with different action spectra are required (Figure 11.6).

1st type : response curve positive in the red and in the blue, negative in the green.

2nd type : response curve positive in the yellow and negative in the blue.

3rd type : response curve positive throughout, corresponding to photopic vision of the human eye at 5500 Å.

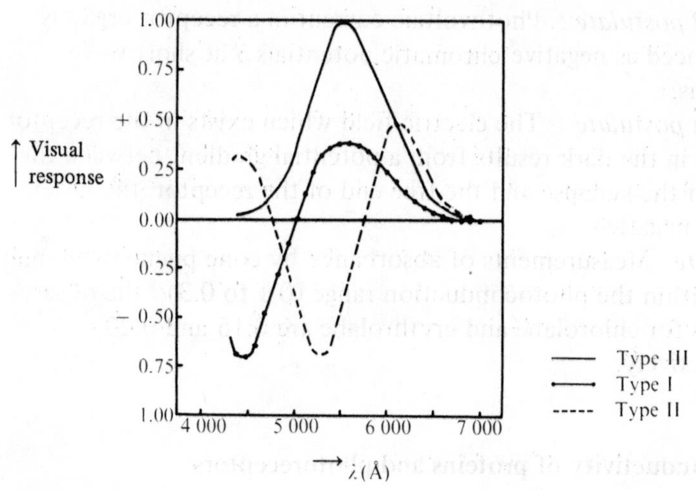

FIGURE 11.6

This third type responds to intensities, while the first two types respond to colors.

A photoconducting system is capable of supplying such a response series and will thus function as a human eye.

B) *Agreement with electrophysiological data*

Experiments carried out on retinas of certain fish showed that there is a close agreement between the potential variation as a function of the wavelength and the variation which is observed in photoconduction experiments.

C) *Postulates of the photoconduction theory of vision (Rosenberg)*

1st postulate : The potentials of the retina originate from a current source located in the external segments of the cones. These currents are due to electronic charge transfer (photoconducting and photovoltaic effects), produced by the absorption of light.

2nd postulate : Photoconduction in a receptor organ is possible owing to the existence of an electric field in the dark. The photoconducting current is evidenced by the positive chromatic potential *S* at long wavelengths.

3rd postulate : Photovoltaic current in a receptor organ is evidenced as negative chromatic potentials S at short wavelengths.

4th postulate : The electric field which exists in the receptor organ in the dark results from a potential gradient between the end of the synapse and the free end of the receptor, the latter being negative.

Note. Measurements of absorbance by cone pigments in man fall within the photoconduction range (0.1 to 0.3): the observed values for chlorolabe and erythrolabe are 0.15 and 0.22, respectively.

6. Conductivity of proteins and photoreceptors

Since the chromophores in the visual pigment are separated, it is reasonable to assume that the electronic charge produced by the absorption of radiation is injected into the bulk of the pigment protein; thus, the question of the conductivity of proteins arises immediately.

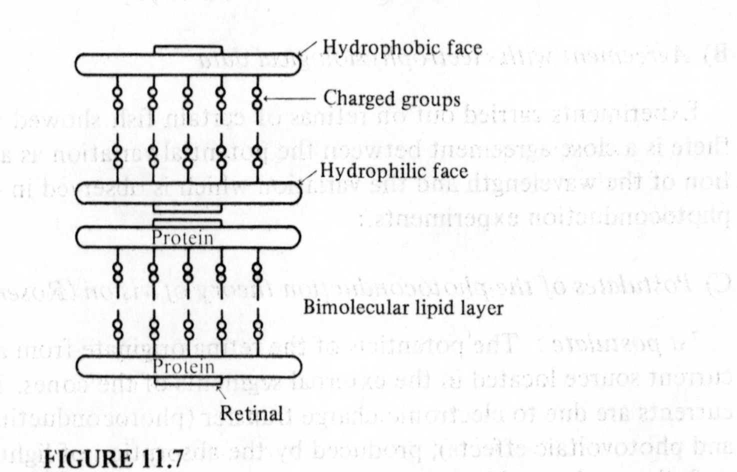

FIGURE 11.7

Crystalline proteins, amorphous proteins, and organic lipoprotein species are semiconductors in the anhydrous state. In the hydrated state, their conductivity becomes much higher, owing to the increase in the value of the dielectric constant of the system. Electronic conductivity predominates only if the degree of hydration is low (below 12%); above this water content the conductivity is ionic or protonic in nature.

In weakly hydrated lamellar systems, electronic conductivity only will be apparent. Visual receptors in fact belong to this class of systems, since the hydrophobic region of the protein alone is free. The hydrophilic region faces the polar lipid layer; the intercalated hydrophobic regions adhere to one another and thus expel water. The hydrophobic retinal will become attached onto the last hydrophobic face of the protein (Figure 11.7).

The characteristic feature of this model is that the water-absorbing sites of the protein are not exposed to water, so that the entire assembly is insoluble; on the other hand, these stacked lamellar structures can be dissociated by detergents. Owing to its position at the hydrophobic extremity of the protein, retinal can slide during the decoloration process.

V. CONCLUSION

These theories, as outlined above, are open to a number of objections. Thus, despite the fact that visual and photosynthetic receptors are highly ordered structures, it is known that they are quite unlike crystalline systems. As concerns semiconduction by pigments in vitro, it has not been proved that activation energies are intrinsic in nature. Moreover, in vivo the corresponding energies could be affected by the biological environment. Donor-acceptor interactions could increase the efficiency of the photoconduction effect. Results obtained by studying crystalline systems have admittedly yielded some very interesting pointers, but the position in reality is probably more complex. In fact, pigment molecules in visual and photosynthetic receptors are organized in lipoprotein membranes, the transfer of an electron across which can involve only one or two molecules. The process seems to take place both by semiconduction and by donor-acceptor interaction. A study of phospholipid membranes with incorporated pigments would seem very promising in this connection

Bibliography

1. Rosenberg, B. Advances in Radiation Biology. Edited by L. G. Augenstein, R. Mason, and M.R. Zelle, 2 : 193–241. Academic Press, N. Y. 1966.
2. Cherry, R. J. – Quarterly Reviews, 22 : 160. 1968.
3. Seliger, H. H. and W. D. McElroy. Light: Physical and Biological Action. Academic Press, N. Y. 1965.
4. Wald, G. Light and Life. Edited by W. D. McElroy and B. Glass, pp. 724–749. John Hopkins Press, Baltimore. 1961.
5. Davydov, A. S. Theory of Molecular Excitons. McGraw-Hill, N. Y. 1962.
6. Castro, G. and J. F. Hornig. – J. Chem. Phys., 42 : 1459. 1965.
7. Geacintov, N. and M. Pope. – J. Chem. Phys., 47 : 1194. 1967.
8. Svaetichin, G. and E. F. McNichol. – Ann. N. Y. Acad. Sci., 74 : 385. 1958.
9. Weisz, S. Z., A. Cobas, P. E. Richardson, H. H. Szmant, and S. Trester. – J. Chem. Phys., 44 : 1364. 1966.

Chapter 12

FLUORESCENCE AND ITS APPLICATIONS IN BIOLOGY

I. INTRODUCTION

Proteins, which are among the most complex substances in biochemistry, fulfill a wide range of biological functions, including enzymatic activity, regulation of hormone activity and immunological activity, and are the main structural factor in the morphology of living organisms. The very large number of different existing proteins is due to the exceedingly large number of possible combinations in which twenty different amino acids can be arranged in linear polypeptide chains. The three-dimensional structure – the so-called tertiary structure – is the result of the interaction of amino acid residues with one another, with the solvent or with the polypeptidic backbone. Generally speaking, hydrophobic amino acid residues are found in the interior of the coiled protein, while the hydrophilic residues are on the outside surface of the protein coil and can interact with the surrounding aqueous medium.

Fluorescence spectroscopy is a technique which is particularly suitable for studying the protein structure. Most proteins contain aromatic amino acids – tryptophan and tyrosine – which fluoresce in UV light. This intrinsic fluorescence – i.e., its intensity, polarization or wavelength – may vary with the changes in the structure of the protein brought about by interaction with small molecules or with other proteins. The number and the sites of the tryptophan and tyrosine residues inside the protein are usually unknown; moreover, the fluorescence of these residues depends on a number of parameters which are difficult to analyze, so that the usual procedure is to employ fluorescent compounds of known structure which serve as detectors ("probes") of changes in the protein structure.

II. DETERMINATION OF COENZYME SITES

It was shown by Theorell and Yonetani that the enzymatic reaction of horse liver alcohol dehydrogenase (LADH) involves an intermediate ternary complex which is formed between the enzyme, the coenzyme, and the substrate. In the same way, inhibitors form a ternary complex with the enzyme and the coenzyme.

The formation of this complex led these workers to determine the number of sites at which reduced diphosphopyridine nucleotide (DPNH) becomes bound to LADH. This enzyme forms strongly fluorescent complexes with DPNH in the presence of excess isobutyramide, whereas LADH or DPNH alone are practically nonfluorescent around 4100 Å. The fluorescence intensity is proportional to the extent of addition of DPNH to the mixture of LADH with isobutyramide, until all the LADH sites are occupied. The fluorescence intensity then remains constant despite further addition of DPNH.

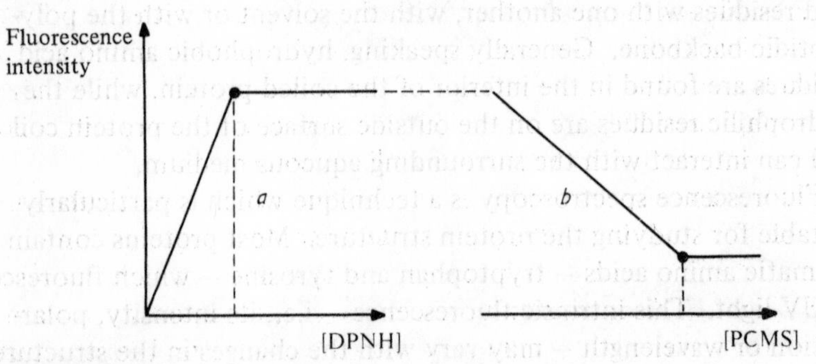

FIGURE 12.1

The experimental curve comprises two straight segments. The abscissa corresponding to their intersection point yields the number of binding sites of LADH for DPNH. This number is two (Figure 12.1a).

Theorell and Bonnichsen showed that DPNH is bound to LADH by way of the SH groups of the enzyme. It is of interest to determine the number of SH groups per LADH molecule. To do this, the LADH-DPNH-isobutyramide ternary complex is titrated against *p*-chloromercuriphenyl sulfonate PCMS, which displaces DPNH from the ternary complex and becomes bound in the complex with one PCMS molecule per SH group. If suitable amounts of PCMS are added to a solution of the ternary complex, its fluorescence decreases in proportion to the amount of PCMS added. In fact, PCMS is not fluorescent and does not quench the fluorescence of DPNH. If PCMS is added in excess, the fluorescence of the mixture attains a minimum steady-state value, which corresponds to the fluorescence of the free DPNH.

The quenching of the fluorescence of the ternary complex brought about by the addition of PCMS is attributed to the liberation of DPNH from the ternary complex. The titration curve (Figure 12.1b) of this complex consists of two intersecting straight lines; the abscissa of the point of intersection corresponds to the number of SH groups per LADH molecule; this number was found to be 28.

III. HYDROPHOBIC FLUORESCENT PROBES

Fluorescent "probes" are compounds, the fluorescent properties of which are modified by a noncovalent interaction with proteins or with other molecules. Several spectroscopic characteristics of the probes, such as emission quantum yield, are strongly affected by the nature of the solvent and by the medium surrounding the macromolecule. For these reasons, they can be used to reveal conformational changes taking place in the proteins.

Such a probe is TNS (2-toluidinylnaphthalene-6-sulfonate), which was used by Edelman and McClure. The compound does not fluoresce in water, but emits a strong fluorescence if bound to a protein. This property is explained by the very high polarity of the first excited state $^1(\pi \rightarrow \pi^*)$ of TNS: the dipole moment of this state is $10D$ times higher than that of the ground state; in a polar solvent, radiationless transitions to the ground state are strongly favored. As a result, the fluorescence quantum yield of TNS decreases in polar solvents.

Studies carried out in the presence of proteins demonstrated that TNS is a hydrophobic probe. When brought into contact with denatured, partly uncoiled proteins, which have their hydrophobic chains exposed to the solvent, the fluorescence of TNS becomes stronger than with native proteins. Bovine serum albumin, which contains hydrophobic bonding sites, increases this fluorescence to a greater extent than, say, lysozyme. The hydrophobic nature of the crevices in the heme in hemoglobin or myoglobin is evidenced by an increased fluorescence quantum yield; in this case the yield is almost unity.

Conformational changes can be detected by fluorescent hydrophobic probes: the native structure of the protein is stabilized by various chain interactions, hydrogen bonds, hydrophobic and electrostatic interactions. If these effects disappear, owing to the influence of the solvent or of the temperature, the protein becomes denatured and its tertiary structure is altered. As a result, some of its hydrophobic residues, which are contained in the interior of the native protein, emerge to the surface where they are exposed to the solvent. Fluorescent probes may be used as very sensitive indicators of denaturation.

Determinations of the number of molecules of a probe bound to a given protein are as yet few. Since the surface of a globular protein is hydrophilic, few probe molecules will be adsorbed on it. A chymotrypsin molecule binds one single TNS molecule. Liver alcohol dehydrogenase binds 2 molecules of 1,8-ANS and bovine serum albumin binds 5 molecules of this compound. These studies are necessary, since they allow the recognition, by comparison, of conformational changes, which are less profound than denaturation and which are reversible.

The study of the conformational change during the passage from chymotrypsinogen to chymotrypsin will serve as an example. Chymotrypsinogen is an enzyme precursor; it has no enzymatic activity, but is converted to chymotrypsin. The activation of chymotrypsinogen can be followed by measuring the activity of the liberated enzyme as a function of time. The conformational change which accompanies this activation may be very accurately observed by following the increase of the fluorescence of TNS incorporated in the mixture.

This increase is exactly parallel to the development of the enzymatic activity. Thus, it is seen that the "probe" responds to the small folding of the polypeptide chain which accompanies the formation of the active site (it is known that this formation is due to the cleavage of a simple peptide bond between arginine in position 15 and the isoleucine in position 16). It could be shown that the TNS was not bound to the active site of the enzyme, but was attached to the protein at a secondary site. This is the more surprising as the active site of chymotrypsin includes a hydrophobic region. Despite the fact that the substrate and the detector are bound to spatially separated sites, the fluorescence of the former changes when substrate analogs become bound to the active site. In the view of the authors, the attachment of the substrate to the site results in a conformational change which no longer permits the TNS to be bound to the enzyme.

Fluorescent probes have many applications. One of the earliest applications, which involves toluidinylnaphthalene sulfonic acid (TNS acid), concerns the effect of an antibiotic – polymyxine – on the bacterial wall. 1-Anilino-8-naphthalene sulfonate (1,8-ANS) has been successfully employed in a study of the swelling of serum albumin at acid pH values. This reagent also served to identify histones of calf thymus. The conformational change of glutamine synthetase, brought about by metal-chelating agents, could be followed by the use of TNS.

Owing to their sensitivity, these probes are particularly useful in studies of antigen-antibody interactions. Interactions between TNS and antibodies to naphthionic acids could be studied in this manner.

We have so far discussed the studies carried out on proteins, but fluorescent detectors can also be utilized in the study of other macromolecules. It was shown by Le Pecq and Paoletti that a trypanocidal dye acted as a fluorescent probe of helicoidal polynucleotides, specific to certain regions of DNA and RNA and was sandwiched between the base pairs.

It is certain that the probe technique is very promising in research on the relationship between the conformation and the function of proteins and of other biological macromolecules.

IV. FLUORESCENCE POLARIZATION

1. Principle

The vibrations of natural light display a symmetry of revolution about the direction of propagation; polarized light has no such symmetry.

Let us consider a system of axes *Ox*, *Oy*, and *Oz*. A fluorescent solution, which is placed at *O*, is irradiated by a beam of polarized light, the electric vector of which is parallel to *Oz*. The fluorescent light is observed along the *Oy* axis, and the intensity of the component I_z which is parallel to *Oz* and of the component I_x which is parallel to *Ox* is determined. The degree of polarization p is given by

$$p = \frac{I_z - I_x}{I_z + I_x}.$$

The fluorescence is polarized if $p \neq 0$, i.e., if $I_z \neq I_x$.

The concept of the "transition moment" of an electronic transition was explained in Chapter 1. The absorption and emission processes are governed by two vectors (transition moments) attached to the molecule; the direction of these vectors mainly depends on the geometry of the molecule. Light absorption is at a maximum

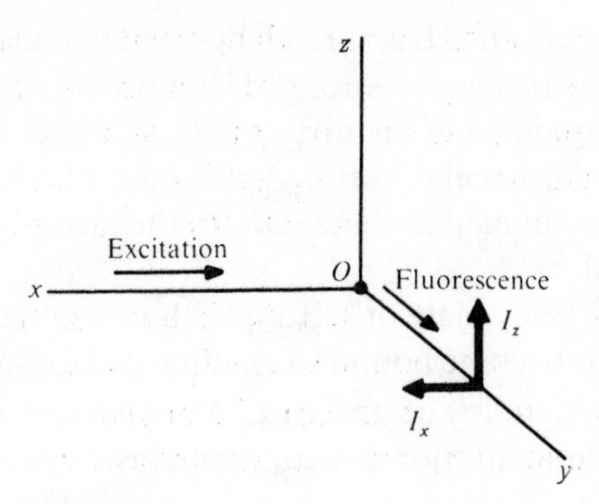

FIGURE 12.2

if the electric light vector is parallel to the transition moment of absorption; even though the molecules in solution are oriented at random, the excited molecules will be similarly oriented. If the excited molecules have enough time to change their orientation prior to the emission, the fluorescence will be very slightly polarized; if, on the contrary, the emission takes place prior to the change in orientation, the fluorescence will be polarized. This state of things will be realized under certain conditions (short lifetime of the excited state, a bulky molecule, a very viscous solvent, etc.).

The following equation gives the degree of polarization p as a function of the angle λ between the transition moments of absorption and emission in a solution excited by polarized light:

$$\frac{1}{p} - \frac{1}{3} = \frac{5}{3} \cdot \left(\frac{2}{3 \cos^2 \lambda - 1} \right).$$

It is seen that p can vary between + 1/2 ($\lambda = 0$; parallel moments of absorption and emission; this is most often the case for $S_0 - S_1$ transitions) and -1/3 (for $\lambda = 90°$).

2. Conjugated proteins

One or more fluorescent groupings may become bound to a protein by a chemical reaction. This reaction should be sufficiently mild not to modify the structure of the protein, and the resulting bond should be sufficiently strong. The most often employed fluorescent groupings include the different isocyanates of fluorescein, Rhodamine B, anthracene, sulfonyl chlorides of anthracene and of 1-dimethylaminonaphthalene (DNS), the latter being the most frequently used.

As a rule, the isocyanate or sulfonyl chloride moieties of fluorescent compounds become bound to α-amino or ϵ-amino groups, and to the sulfhydryl groups of proteins. The physical constants of the proteins (sedimentation constant, intrinsic viscosity) remain unaffected.

3. Determination of molecular volumes and molecular asymmetries

Perrin's equation describes the variations in fluorescence polarization as a function of the temperature T and viscosity η of the solution for a spherical molecule excited by polarized light:

$$\frac{1}{p} = \frac{1}{p_0} + \left(\frac{1}{p_0} - \frac{1}{3}\right) \frac{R\tau}{V} \frac{T}{\eta}.$$

If the variation of $1/p$ is measured as a function of T/η, the hydrodynamic molecular volume V can be determined. The molecular weight of the protein can be deduced from V, if its degree of hydration in solution is known.

This case, i.e. a homogeneous population of spherical molecules, is only rarely encountered in practice; a better model of proteins is an ellipsoid, the eccentricity of which represents the asymmetry of the molecule.

If, at a constant volume, the eccentricity increases, fluorescence polarization increases as well, and the curve $1/p = f(T/\eta)$ approaches the T/η axis (i.e., the slope decreases). In the case of prolate ellipsoids, the rectilinear character of the curve is not much changed when the slope becomes less steep; the same applies to oblate ellipsoids, the eccentricity of which is less than 3. If the eccentricity is larger, the curvature of the plot increases and the curve becomes concave toward the T/η axis.

When there is no molecular asymmetry, the plot is convex toward the T/η axis, if the protein is heterogeneous with respect to its molecular volume. This effect may become superposed onto the curvature which is due to the asymmetry. The homogeneity of the protein must then be tested by other techniques.

4. Other applications of fluorescence polarization

A) *Determination of the lifetime of the excited state*

If the volume V is known, the lifetime τ may be determined from Perrin's equation.

B) *Energy transfer in biological systems*

The weak polarization of the fluorescence emitted by chlorophyll in chloroplasts is one of the most powerful arguments in favor of an energy transfer taking place in chloroplasts. In proteins, energy transfer between tyrosine residues is probably responsible for the smaller values of p_0 which are observed in tyrosine-containing proteins as compared to phenol or tyrosine.

C) *Study of interaction of fluorescent substances with proteins*

When a fluorescent substance interacts with a protein by ionic or or by secondary forces, its fluorescence becomes polarized, since its rotation is more or less restricted to that of the protein. If the concentration ratio between the protein and the fluorescent substance is increased, the fluorescent compound can be totally bound to the protein. To determine the fluorescence of this complex, a series of solutions of a low constant concentration in the fluorescent compound and of progressively increasing concentrations in protein are prepared. The polarized fluorescence of the substance entirely bound to the protein is quenched by the nonpolarized fluorescence of the free substance. In this way the fraction of the fluorescent substance bound to the protein can be found.

The dissociation equilibrium constant K of the complex between protein and the fluorescent substance, and also the number n of the absorption sites could be obtained by this method for the complex of serum albumin with fluorescein or with naphthylamine sulfonic acids.

A particularly important application is the study of enzyme-coenzyme complexes.

Velick studied the enzyme-coenzyme complexes of glyceraldehyde-3-phosphate dehydrogenase (GPD) and of lactic acid dehydrogenase (LDH); he could determine the number of sites of fixation of DPN or DPNH on GPD.

A sample containing $6 \cdot 10^{-9}$ mole of DPNH in 1 cm^3 of buffer solution (pH 7) is titrated against a solution of the GPD enzyme. The fluorescence intensity of DPNH decreases and its polarization increases as the enzyme is added (Figure 12.3).

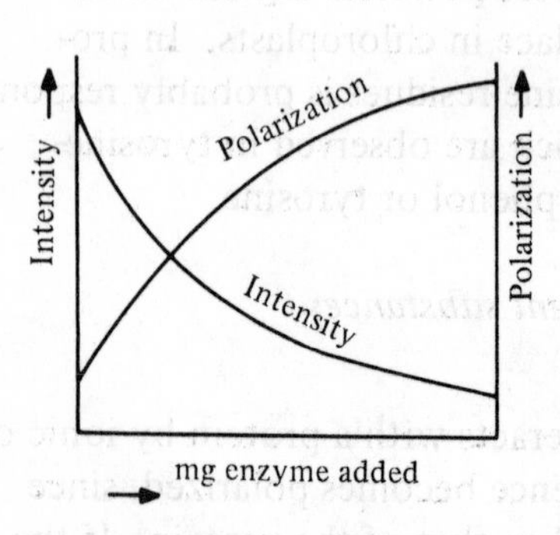

FIGURE 12.3

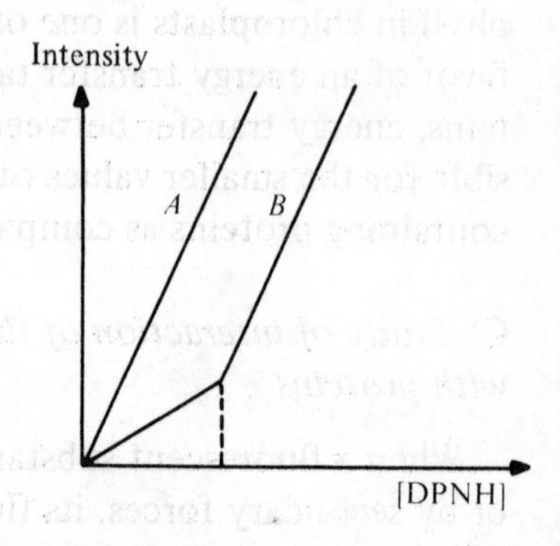

FIGURE 12.4

The stoichiometry of the GPD-DPNH complex is determined by titrating the enzyme against DPNH (Figure 12.4).

The comparison curve A is that given by a solution of DPNH not containing the enzyme. Curve B shows the fluorescence intensity of the mixture; it includes two intersecting rectilinear segments, the intersection point of which corresponds to the enzyme-coenzyme complex. The abscissa of this point gives the stoichiometry of the complex: GPD $(DPNH)_3$.

Velick also studied the competitive fixation of DPNH and DPN on GPD by titrating the previously prepared complex against DPN; his curves are shown in Figure 12.5. DPN was not fluorescent in the experimental range.

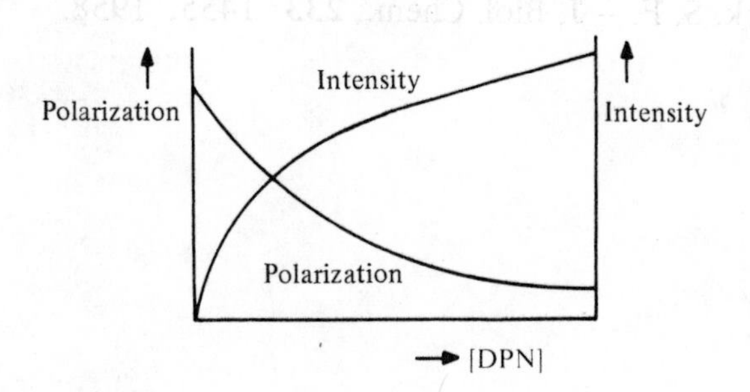

FIGURE 12.5

Velick also measured the ratios of equilibrium constants of the dissociation of GPD-DPNH complexes and showed that DPN is bound more readily than DPNH. DPN displaces the fixed DPNH; this is the course of an enzymatic reaction in which two forms of the coenzyme occupy the same sites.

5. General comments

The fluorescence polarization method is less sensitive to variations in pH, salt concentration, and protein concentration than other methods (rate of diffusion, rotatory power, dielectric constant, viscosity). It is very sensitive, is applicable within a fairly wide range of temperatures, and is suitable for the study of conjugated proteins in the presence of other, nonfluorescent proteins. The technique is simple, and the apparatus required is not expensive.

Bibliography

1. Edelman, G. M. and W. O. McClure. – Accounts of Chemical Research, **1** : 65. 1968.
2. Theorell, M. and T. Yonetani. – Arch. Bioch. Biophys., **99** : 433. 1962; Ibid., Suppl. **1** : 209. 1962.
3. Theorell, M. and R. Bonnichsen. – Acta Chem. Scand., **5** : 1105. 1951.
4. Velick, S. F. – J. Biol. Chem., **233** : 1455. 1958.

Chapter 13

CHARGE TRANSFER COMPLEXES

I. PRINCIPLE

Charge transfer complexes (see Chapter 3) are generated during a more or less complete transfer of an electron from a donor to an acceptor. According to Mulliken, these complexes have a resonance structure: the interaction of donor D with the acceptor A can be described by saying that when D and A form a complex, its wave function may be written as follows:

$$\psi_N = a\psi_{(DA)} + b\psi_{(D^+A^-)} \qquad a > b \quad \text{(ground state)},$$

$$\psi_E = b^* \psi_{(DA)} - a^* \psi_{(D^+A^-)} \qquad a^* > b^* \quad \text{(excited state)}.$$

DA represents a structure in which D and A are interconnected solely by classical bonding forces (electrostatic forces, hydrogen bonds, etc.); D^+A^- represents a structure in which an electron has been transferred from D to A; this bond is then present in addition to the other interaction forces of DA.

The diagram in Figure 13.1 represents the different states and energy relationships for a charge transfer complex according to Mulliken's theory.

The transition frequency of charge transfer is

$$h\nu = W_E - W_N = I_D^v - E_A^v - (G_1 - G_0) + (X_1 - X_0)\,.$$

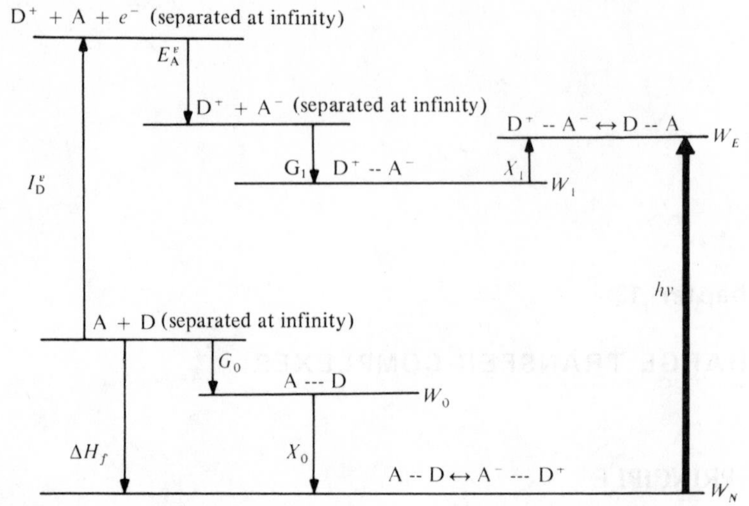

I_D^v = vertical ionization potential of D.
E_A^v = vertical electron affinity of A.
ΔH_f = heat of formation of the complex in the ground state.
G_0 = energy of classical interaction forces between A and D in the ground state
G_1 = energy of classical interaction forces between D^+ and A^- in the excited sta
X_0 = resonance energy between A--D and A^-D^+ in the ground state.
X_1 = resonance energy between A--D and A^-D^+ in the excited state.

FIGURE 13.1

Since most π, π^* complexes are weak, $a \gg b$ and $b^* \gg a^*$, the contribution of $(X_1 - X_0)$ is small enough to be neglected and we may write

$$h\nu = I_D^v - E_A^v - W$$

where $W = G_1 - G_0$ is the dissociation energy of the excited state of the complex.

There is a linear relationship between I_D^v and $h\nu$ for charge transfer complexes between a common acceptor and similar donors (in such a case W may be assumed constant). There is also a linear relationship between E_D^v and $h\nu$ if a common donor and different similar acceptors are employed. These linear relationships can be utilized to determine electron affinities or ionization potentials.

II. COMPLEXES IN BIOCHEMISTRY

Various authors have recently postulated the formation of charge transfer complexes between molecules of biological interest, in particular for conjugated biomolecules which are expected to have low ionization potentials and strong electron affinities. These workers considered the part played by these complexes in the mechanism of biochemical reactions and in the structure of certain cell components (nucleic acids, mitochondria, quantosomes). The following biomolecules are involved:

a) the principal coenzymes of redox processes, in particular the nicotinamide cycle of pyridine nucleotides and the isoalloxazine cycle of flavines;

b) purines;

c) indole compounds: tryptophan, serotonin;

d) quinones.

The determination of the donor–acceptor properties of biomolecules is particularly important. Since charge transfer transitions are rarely observed in biomolecules, these determinations were carried out by various techniques.

1. Method of molecular orbitals (Hueckel's approximation)

Calculations yield energy values in the form

$$E_i = \alpha + K_i \beta ,$$

where α and β are resonance integrals. The values of K_i lie between 0 and 15 for the highest occupied molecular orbitals and between -1.5 and 0 for the lowest vacant orbitals. As the K_i value approaches zero, the electron-donating or electron-accepting properties of the molecule become more pronounced. Tables of molecules participating in biological processes have been compiled.

2. Ionization potential method

The value of the coefficient K_i can be obtained from the values of the ionization potential I from the relationship:

$$I(\text{eV}) = (3.14 \pm 0.24)\,K_i + (6.24 \pm 0.10).$$

3. Polarographic method

A satisfactory correlation can also be established between the value of K_i and the electrochemical behavior of the compound. Studies of the electron-donating and electron-accepting properties of purines and pyrimidines showed that there is a perfect correlation between their K_i-values and the readiness with which these molecules are polarographically oxidized or reduced. Uric acid ($K_i = 0.17$) is the most readily oxidized compound studied; guanine ($K_i = 0.30$) is the most readily oxidizable nucleic base studied.

4. Classification of biomolecules

Following these determinations, biomolecules can be subdivided into three groups:

a) compounds which act mainly as donors: purines, pyrimidines, proteinic α-amino acids, reduced forms of flavine and pyridine nucleotides. Certain compounds of interest as potential drugs, such as phenothiazines, are very good donors;

b) compounds which act mainly as acceptors: oxidized forms of flavine and pyridine nucleotides, certain pteridines, quinones, and bile pigments;

c) compounds which act both as donors and acceptors: porphyrins, carotenes, retinenes.

Shifrin studied compounds modeling the enzyme-coenzyme interaction of the type shown below, in which an aromatic donor nucleus undergoes exchange reactions with an acceptor residue; in the diagram, AAA represents the conjugated aromatic nucleus of the amino acid.

CH_2—CH_2 $Cl^{\ominus}$
$N^{\oplus}$
AAA
$CONH_2$

There is a correlation between the frequency of the charge transfer band and the ionization potential of the aromatic donor (Figure 12.2). These studies indicate that indole and indole derivatives (tryptophan) are strong donors; such a correlation could not be established for imidazole.

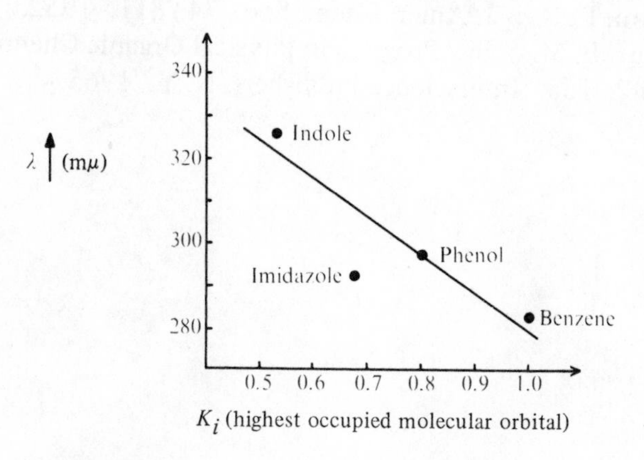

FIGURE 13.2

We must now inquire into the role played by charge transfers in the formation of molecular complexes.

Calculations showed that the contribution of charge transfer forces to the energy of autoassociation of purines or pyrimidines in aqueous solution is quite small and amounts to only a fraction of the van der Waals–London forces. On the other hand, the contraction of the interplanar distance in the crystals of 8-azaguanine may be due to a charge transfer interaction. Experimental facts observed for nucleic acids could also be interpreted in terms of the semiconductivity of these molecules. It is always dangerous to assume that a biological phenomenon involves charge transfer complexes, unless there is solid evidence to support such an assumption.

Bibliography

1. Pullman, A. and B. Pullman. – In: Quantum Theory of Atoms, Molecules, Solid State, pp.345–359. Academic Press Inc., N. Y. 1966.
2. Mulliken, R. S. – J. Amer. Chem. Soc., **74** : 811. 1952.
3. Kosower, E. M. – In: Progress in Physical Organic Chemistry, 3 : 87–163. Interscience Publishers, N. Y. 1965.

Chapter 14

EXPERIMENTAL TECHNIQUES OF PHOTOCHEMISTRY

I. APPARATUS

Any irradiation system comprises a source of radiation, a filtering system and a cell with the substance to be irradiated, which can be solid, liquid, or gaseous. Many other accessories can also be utilized: lenses, stops, cell thermostating unit, photosensitive cell, etc.

1. Radiation sources

The choice of the radiation source will obviously depend on the wavelength of the light which is absorbed by the irradiated substance. Most lamps utilized in photochemistry are mercury vapor lamps. Individual lamps show significant differences in their power and in the nature of the emitted spectrum according to the particular value of the metallic vapor pressure, foreign gases which may or may not be present, the nature of the envelope, etc. For a list of commercially available lamps see reference /2/ in Chapter 4.

A) *Low pressure mercury arc lamps (resonance lamps)*

In such lamps the pressure of mercury is 0.005 to 0.1 mm. Two main lines are emitted: 2537 Å and 1849 Å. If the lamp envelope is not made of a highly transparent brand of quartz, all 1849 Å radiation will be absorbed by the envelope and the lamp will only emit the 2537 Å radiation (90% of total intensity). If it is desired to work with 1849 Å radiation, the lamp envelope must be made of Suprasil or of a similar brand of quartz glass. Low pressure lamps work at low current intensities, under fairly high voltages and their power demand is low (less than 100 W); their lifetimes are long (about 10,000 hours).

B) *Medium pressure mercury arc lamps*

In these lamps the pressure of mercury is about one atmosphere. The lamps emit numerous lines in the UV and in the visible, the main lines being 3650-3663, 3126-3132, 3022-2967, 2652-2655, and 2537-2380 Å. For a fixed size, the UV energy is much higher than in low pressure arc lamps; these lamps can accordingly be used in conjunction with filters or monochromators.

The power demand of these lamps can be very high (several thousands of watts) and their useful service life usually does not exceed 1000 hours.

C) *High pressure mercury arc lamps*

The pressure of mercury in these lamps is more than 10 atm. These lamps are the most powerful sources of UV radiation, but their emission mostly lies beyond 3000 Å. The spectroscopic lines widen if the pressure and the temperature of the lamp are raised; the emission of ultra-high pressure (several hundreds of atmospheres) lamps is almost continuous. The emission in the visible is very important. These lamps are operated at very high temperatures and must be cooled. High pressure lamps which are near-point sources (mercury, xenon, mercury-xenon arcs) are also available commercially; they have longer lifetimes than ordinary high pressure lamps and are extensively used in determinations of quantum yields.

The power demand of lamps of this kind may be as high as 400,000 W. Their useful service life is quite short (100–200 hrs).

D) *Sodium vapor lamps*

Low-pressure sodium vapor lamps emit an almost monochromatic 5890 Å radiation. High pressure lamps emit several fairly wide lines in the visible.

E) *Discharge lamps for the far UV*

These are low-pressure lamps (0.1–10 mm Hg). The characteristic emission of the gas with which the lamp is filled is almost monochromatic: iodine 2060 Å, krypton 1165-1236 Å, xenon 1296 and 1470 Å, hydrogen 1216 Å, argon 1067-1048 Å, neon 744-736 Å, helium 584 Å.

F) *Lasers*

Lasers have not yet been much used in photochemistry, but their future in this field seems assured; lasers produce a coherent light beam which is very powerful and monochromatic. Lasers emitting in the UV are as yet few, but we may note a pulsed nitrogen laser emitting at 3371 Å, and a neodymium laser which emits at 2645 Å after two frequency doublings.

2. Filters

There are four main kinds of filter systems: glass filters, chemical filters, interference filters, and monochromators.

Different glasses absorb radiation in different ways. Very pure quartz glass transmits the 1849 Å mercury line. In the farther UV calcium fluoride, lithium fluoride or sapphire must be used. Pyrex glass is practically opaque to light with wavelengths below 2900 Å. Many brands of glass filters are commercially available, Vycor and Corning are the best known ones. Some glasses transmit all wavelengths longer than a certain limiting value; others transmit a band of light comprised between two limiting wavelengths.

Chemical filters consist of a glass cell containing a chemical compound or, more often, a solution of this compound. Accordingly, these systems have two filters: the glass and the chemical compound.

If several substances are combined, fairly narrow transmission bands can be obtained and one or more emission lines of a lamp thus isolated. For descriptions of a large number of chemical filters the reader is referred to the treatise by Calvert and Pitts. The main disadvantage of chemical filters is their poor stability. After the filter has been irradiated for a certain period of time, its percentage transmission changes, and special care must be taken in quantitative work.

Interference filters yield very narrow transmission bands, but their percentage transmission is often low. These filters are usually sensitive to heat and must be protected from the heat of the lamp by interposing a glass plate or a cell filled with distilled water between the lamp and the filter.

Two types of monochromators exist: diffraction (grating) monochromator and refraction (prism) monochromator. They are particularly well suited to isolating monochromatic radiation, especially in mid-UV region (2000–3000 Å), in which other filtering systems are usually inefficient. Another advantage of monochromators is that they are not perishable (especially prism monochromators), contrary to other filtering systems. They have the disadvantage of a low transmission; this does not apply to certain models which have been specially constructed for the photochemist.

A few remarks concerning the utilization of these systems may not be out of place. The first thing to do is to determine the width of the transmission band of the filter and the percentage of the light which is transmitted in this range; these measurements must be periodically repeated in order to reveal any aging of the filter (this especially applies to colored glasses, chemical filters, and interference filters). The percent transmission may be determined with the aid of a photoelectric cell connected to a galvanometer; the standard transmission is determined without the filter, after which the filter is inserted and the difference measured. If a monochromator is employed, it is undesirable to open the window in order to raise the radiation intensity, since then the transmission band becomes wider. The spectrum emitted by the lamp must be studied in order to ensure that lines situated near the desired line are not transmitted together with it. For transmission curves of the various filtering systems the reader is referred to the books by Calvert and Pitts or by Schönberg, and to the catalogs of the manufacturing companies.

3. Cells

The cell chosen must be suited to the type of irradiation employed (preparative or analytical, qualitative or quantitative), to the nature of the irradiated substance (solid, liquid, gas), to the radiation wavelength (quartz, Pyrex, etc. cells) and to the optical assembly used (optical train, immersion system, etc.).

A) *Optical train* (Figure 14.1)

(1) Lamp
(2) Lens system
(3) Stop
(4) Filter system
(5) Cell
(6) Photoelectric cell

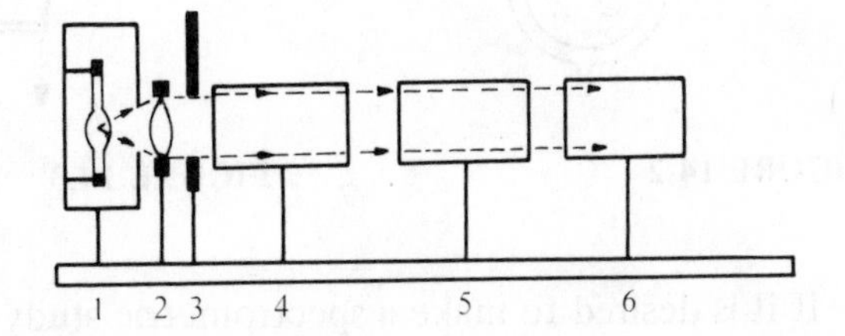

FIGURE 14.1

Such a system is specially suited to quantitative work. If the different parts are mounted on an optical bench, the determinations are highly reproducible. Its disadvantage is that only a small part of the radiation emitted by the lamp is utilized, and for this reason it will not be often used in preparative work.

In quantitative experiments, the cells must have two plane-parallel windows (preferably circular in shape), and the incident radiation beam must be perpendicular to the plane of the windows. The material of which the cell windows are made will vary with the wavelength. The system is suitable for working with solids, liquids, or gases.

The irradiation cell may be provided with a jacket for the circulation of the thermostating liquid. The windows of the jacket must be parallel to the windows of the cell. The thermostating liquid chosen must be transparent to the wavelength of the radiation. If the study is conducted on a vitrified solution, the jacket can be of the Dewar type and the thermostating liquid may be liquid nitrogen (Figure 14.2).

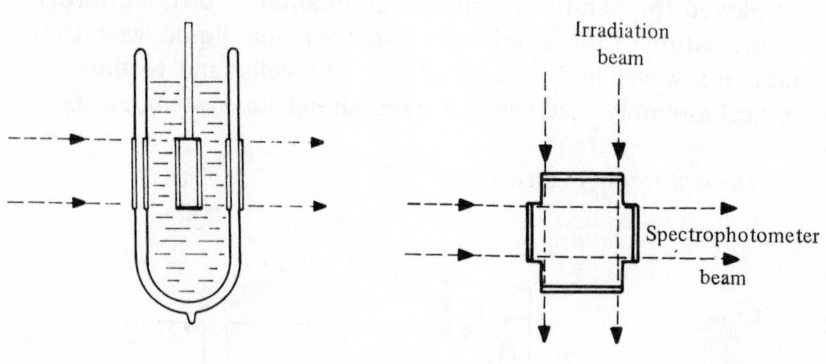

FIGURE 14.2 **FIGURE 14.3**

If it is desired to make a spectrometric study of the changes taking place in the irradiated substance, the cells used may have two additional windows perpendicular to the other pair of windows situated in the path of the spectrophotometer beam. The irradiation is then carried out in the spectrophotometer. Figure 14.3 gives a schematic representation of this arrangement. EPR measurements can also be carried out during the irradiation.

B) *Immersion system*

For a lamp to function at its maximum efficiency, the irradiated substance must be able to absorb light emitted in all directions; this means that it must surround the lamp. Many systems of immersion lamps are utilized in preparative photochemistry; some of them are commercially available. Such a system is schematically represented in Figure 14.4.

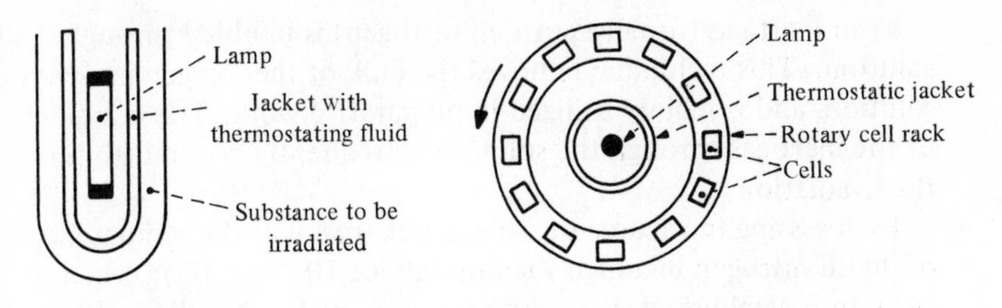

FIGURE 14.4

FIGURE 14.5

The walls of the jacket around the lamp together with the thermostating fluid also act as filters. Other filtering jackets may also be interposed between the lamp and the substance to be irradiated.

Quantitative determinations are usually not possible in such systems, because the intensity of the light emitted by the lamp is not identical in all directions. This disadvantage may be obviated if the system represented in Figure 14.5 is employed.

The lamp is permanently immersed in a filtering thermostatic jacket. This jacket is placed in the center of a circular rack with a certain number of identical cells, with parallel faces; when the rack is rotated, we may assume that all the cells are uniformly irradiated. This system is particularly suitable if all the samples must receive the same amount of light (inhibition experiments, etc.); the amount of the light received by each cell can be determined by actinometry.

II. PHOTOCHEMICAL TECHNIQUES

1. Degassing of solutions

The presence of the smallest amounts of oxygen may bring about major variations in the course of a photochemical reaction; it is therefore important that the solution be degassed as thoroughly as possible. Two techniques can be employed for this purpose:

a) an inert gas (usually purified nitrogen) is bubbled through the solution. This technique removes the bulk of the oxygen present in solution, and is mainly utilized in preparative work. The bubbling of the inert gas through the solution is frequently continued during the irradiation;

b) degassing in vacuo: the solution is frozen at the temperature of liquid nitrogen in a high vacuum (about 10^{-6} mm Hg); when the vacuum is established, the solution is removed and is allowed to return to room temperature. After several such cycles (3–4 are usually sufficient), the degree of degassing becomes very high. It will be noted that a solution must never be degassed in a quartz cell; the degassing is effected in a Pyrex cell and the sample is then transferred to a connecting quartz cell in which it is irradiated after the assembly has been sealed in vacuo.

2. Actinometry

The measurement of the radiation intensity is unavoidable in determinations of quantum yields. These determinations can be absolute (thermocouple) or relative (chemical actinometry). Organic chemists carry out relative determinations only.

A) *Different types of actinometers*

The quantum yield of a transformation which takes place in a good chemical actinometer under irradiation should not significantly change the experimental conditions (temperature, concentration, light intensity, and wavelength within a certain range); the analysis of the transformation thus produced must be simple and accurate. There are a number of systems which satisfy these conditions fairly well; the reader is referred to the treatise by Calvert and Pitts.

a) *Vapor phase actinometry*

2500–3200 Å: acetone or diethylketone,

1800–2500 Å: hydrobromic acid.

Other systems have been used in the far UV: photolysis of oxygen, of nitrous oxide, of CO_2.

b) *Solid phase actinometry*

The only system currently in use is the isomerization of *o*-nitrobenzaldehyde to *o*-nitrosobenzoic acid (Chapter 4) in a solid medium (a KBr capsule or a polymer film).

c) *Liquid phase actinometry*

The most frequently used system is the potassium ferrioxalate actinometer of Hatchard and Parker. Its effective range is very wide: between 2500 and 4800 Å; if certain precautions are taken, it may even be used up to 5770 Å.

The useful range of uranyl oxalate is between 2080 and 4350 Å; this compound is also frequently used, but is much less sensitive than potassium ferrioxalate.

B) *Actinometric determination of the intensity of the absorbed radiation*

Substance *A*, which is utilized as actinometer, is irradiated for a given length of time under the same conditions as the sample to be irradiated (same cell, same optical assembly, etc.). The light intensity absorbed per unit time will be

$$I_0^i = \frac{n}{\Phi_A . t . (1 - 10^{-\varepsilon[A]l})} \text{ quanta} \cdot \text{sec}^{-1},$$

where n is the number of molecules formed during the time of irradiation t, and Φ_A is the absolute quantum yield of the product, determined with the aid of a thermocouple. Its value is known for classical actinometers. The expression $(1 - 10^{-\varepsilon[A]l})$ is the fraction of the incident light absorbed by the actinometer. This value can be measured directly in a photoelectric cell, or else may be calculated from the known values of concentration $[A]$, extinction coefficient ϵ, and the cell length l. I_0^i is the intensity of the light beam on the inner side of the first window of the cell. Allowance may also have to be made for the reflection of light from cell windows.

C) *Determination of the absorbed fraction of the incident light*

If an optical bench is employed, a photocell is placed behind the irradiation cell, in order to determine the fraction of the incident light absorbed by the solution.

The first step is to determine the deflection Δ_0 of the galvanometer with an empty cell, and then its deflection Δ which is obtained when the cell is filled with the solution to be irradiated. Since the two faces of the cell also reflect light, the fraction of the light absorbed by the solution is not $(1 - \Delta/\Delta_0)$ but

$$\frac{I_a}{I_0^i} = \left(1 - \frac{\Delta}{\Delta_0}\right) \cdot F,$$

where I_a is the real absorbed intensity, I_0^i is the intensity of the radiation arriving at the interior of the first cell window (as before), and F is a factor accounting for the different reflections from the cell windows.

If the solution absorbs all the incident light, $F = 1$. The reader will find a table of the approximate values of F for a quartz cell, different wavelengths and different percentage absorption in the treatise by Calvert and Pitts (p.794).

In determining the quantum yield of a reaction, these difficulties are best avoided by using identical percentage transmissions for the irradiated and for the actinometric solutions. Under these conditions, the exact percentage transmission need not be known. The quantum yield Φ of the reaction will be:

$$\Phi = \frac{n \cdot t_A \cdot \Phi_A}{n_A \cdot t},$$

where t_A is the duration of irradiation of the actinometer, n_A is the number of molecules formed as a result of irradiation of the actinometer, Φ_A is the quantum yield of the actinometer, t is the time of irradiation of the solution, and n is the number of molecules formed during the time t.

3. Flash photolysis

A very brief flash of very strong light may result in the formation of large concentrations of intermediate species (free radicals, triplets, unstable molecules). These intermediate species can be detected by spectroscopic methods, so that the primary processes of certain photochemical reactions can be studied directly.

Radiation sources. Various types of lamps can be used, some of which are now on the market:

a) electric discharge between inert metal electrodes in a gas under a low pressure. This type is the most commonly used. Some lamps emit very high energy flashes (up to 33,600) joules) during about 0.1 millisecond. Others have much lower energies and much shorter emission times (0.0016 joule and 10^{-9} sec);

b) high power spark between magnesium electrodes or volatile metal electrodes. The duration of the flash is 10^{-5} – 10^{-3} sec, and the energy may be as high as 400,000 joules. The emission is more highly monochromatic than in a), which is an advantage;

c) high intensity flash produced by an exploding metal wire. A 0.3 millisecond flash; energy 1056 joules;

d) lasers. Linqvist utilized a pulsed nitrogen laser, with 10^{-8} sec pulses, emitting at 3371 Å, but its energy was only 10^{-4} joules. A neodymium laser (p.245) is much more powerful (0.1 joule).

4. Vapor phase photolysis

Irradiations in the vapor phase involve special techniques and a more complex apparatus than that utilized in liquid phase irradiations. The irradiation cell must be maintained at preset, rigorously constant temperatures and pressures. The volatile reaction products are channeled through a system of stopcocks and are condensed by selective cooling. An analysis unit (vapor phase chromatography and/or mass spectrometry) is usually included in the apparatus and serves to make a direct examination of the reaction products.

5. Solid phase photochemistry

We have already discussed the irradiation of solutions vitrified at a low temperature (Figure 14.4). Other kinds of solid phase irradiation include irradiation of crystals, in the pure form or in a KBr matrix, and irradiation of polymer films. In this last case, the photochemistry of the polymer is studied; if the polymer is radiation-resistant, the photochemistry of a substance incorporated into the film may be investigated.

Bibliography

1. Calvert, J. G. and J. N. Pitts, Jr. Photochemistry. John Wiley and Sons, Inc., N. Y. 1966; Schönberg, A. Preparative Organic Photochemistry. Springer-Verlag, Berlin. 1968.
2. For techniques of luminescence which were not studied in this chapter: Hercules, D. M. Fluorescence and Phosphorescence Analysis. Interscience Publishers, N. Y. 1966; Parker, C. A. Advances in Photochemistry. Edited by W.A. Noyes, Jr., G. S. Hammond, and J. N. Pitts, Jr., 2 : 305–383, Interscience Publishers, N. Y. 1964.

INDEX

ABOUT THIS BOOK

An excellent primer for organic chemists and biochemists, presenting the first principles of photochemistry without recourse to an unduly rigorous mathematical treatment. This is followed by highly interesting accounts of a number of selected topics, both "classical" (e.g., photochemistry of carbonyl compounds) and very recent (electrocyclic reactions, the visual process, applications of fluorescence in biology, etc.). A short account of the experimental techniques and the apparatus employed forms the final chapter of the book.

PARTIAL TABLE OF CONTENTS: Laws of photochemistry; electronic transitions; photochemical reactions; photochemistry of benzene; photochemistry of aldehydes and ketones; photosensitized reactions; photochromism; chemiluminescence and bioluminescence; the visual process; applications of fluorescence in biology; experimental techniques used in photochemistry.